# pesticide residues in food – 1977

report of the joint meeting

of the

## fao panel of experts on pesticide residues and environment

and the

## who expert committee on pesticide residues

held in geneva, 6-15 december 1977

**FOOD AND AGRICULTURE ORGANIZATION OF THE UNITED NATIONS**
Rome 1978

Monographs containing further details of the toxicological evaluations and summaries of the evidence upon which the recommendations for maximum residue limits are based, are issued by FAO under the title:

Pesticide Residues in Food – 1977 Evaluations (FAO Plant Production and Protection Paper 10 Sup.)

Copies can be obtained from the Distribution and Sales Section, FAO, Via delle Terme di Caracalla, Rome 00100, Italy or through authorized FAO Sales Agents and Booksellers.

M-84

ISBN 92-5-100578-8

1977 JOINT MEETING OF THE FAO PANEL OF EXPERTS ON
PESTICIDE RESIDUES AND ENVIRONMENT AND THE
WHO EXPERT COMMITTEE ON PESTICIDE RESIDUES

Geneva, 6-15 December 1977

Members of the FAO Panel of Experts on Pesticide Residues and the Environment

Dr. D.C. Abbott, Deputy Director (Customer Services), Laboratory of the Government
Chemist, Department of Industry, London, England

Dr. A.F.H. Besemer, Head, Pesticides Division, Plant Protection Service, Ministry
of Agriculture, and Agricultural University (Phytopharmacy), Wageningen, Netherlands
(Vice-Chairman)

Professor G. Bressau, Head, Pesticides Unit, Division of Food Chemistry, Federal
Health Office, D 1000 Berlin 33 (West)

Mr. J.G. Cummings, Chief, Chemistry Branch, Registration Division, Office of Pesticide
Programmes, Environmental Protection Agency, Washington, D.C., USA

Dr. E.D. Magallona, Assistant Professor and Head, Pesticide Residue Laboratory,
University of the Philippines at Los Baños, College, Laguna, Philippines

Mr. J.T. Snelson, Pesticides Coordinator, Department of Primary Industry, Canberra,
Australia (Rapporteur)

Observer Invited by FAO

Mr. A.J. Pieters, Chairman, Codex Committee on Pesticide Residues, Directorate of
Public Health, Foodstuffs Division, Leidschendam, Netherlands

Members of the WHO Expert Committee on Pesticide Residues

Dr. W.F. Almeida, Director, Division of Animal Biology, Biological Institute, Sao
Paolo, Brazil

Mr. D.J. Clegg, Head, Pesticide Section, Toxicological Evaluation Division, Foods
Directorate, Health Protection Branch, Department of National Health and Welfare,
Tunney's Pasture, Ottawa, K1A 0L2, Canada (Chairman)

Professor F. Coulston, Director, Institute of Comparative and Human Toxicology,
Albany Medical College, Union University, Albany, N.Y., USA

Dr. M. van Logten, Head, Laboratory for General Toxicology, National Institute of Public Health, Bilthoven, Netherlands

Professor E. Mahboubi, School of Public Health and Institute of Public Health Research, University of Teheran, Teheran, Iran [1]

Professor D.V. Parke, Chairman, Department of Biochemistry, University of Surrey, Guildford, England (Rapporteur)


Secretariat

Dr. C. Agthe, Chief, Food Safety Unit, WHO, Geneva, Switzerland

Dr. L. Brader, Chief, Plant Protection Service, FAO, Rome, Italy

Dr. A. Hassan, Chemical Residue and Pollution Section, Joint FAO/IAEA Division of Atomic Energy in Agriculture, Vienna, Austria

Dr. J.E. Huff, Unit of Chemical Carcinogenesis, International Agency for Research on Cancer, Lyon, France

Dr. L. Ladomery, Food Standards Officer, Joint FAO/WHO Food Standards Programme, FAO, Rome, Italy

Mr. A.F. Machin, Senior Research Officer, Central Veterinary Laboratory, Ministry of Agriculture, Fisheries and Food, New Haw, Weybridge, Surrey KT15 3NB, England (Consultant)

Dr. Orville E. Paynter, Department of Commerce, Office of Environmental Affairs, Washington, D.C., USA (Temporary Advisor)

Mr. N. Saito, Plant Protection Service, FAO, Rome, Italy

Professor C. Schlatter, Institute of Toxicology, Federal Institute of Technology and University of Zürich, CH-8603 Schwerzenbach/Zürich, Switzerland (Temporary Advisor)

Dr. E.E. Turtle, Pesticides Specialist, Plant Protection Service, FAO, Rome, Italy (Joint Secretary)

Dr. G. Vettorazzi, Scientist, Food Safety Unit, WHO, Geneva, Switzerland (Joint Secretary)

---

[1] Present address: The Eppley Institute for Research in Cancer, The University of Nebraska College of Medicine, Omaha, Nebraska, USA

CONTENTS

CONTENTS cont.

## CONTENTS cont.



(Note:  Compounds previously evaluated are indicated by *)

W-84 ISBN 92-5-100578-8

PESTICIDE RESIDUES IN FOOD

Report of the 1977 Joint FAO/WHO Meeting

OPENING STATEMENT

A Joint Meeting of the FAO Panel of Experts on Pesticide Residues and the Environment and the WHO Expert Committee on Pesticide Residues was held in Geneva from 6 to 15 December 1977.  The Meeting was opened by Dr. D. Tejada-de-Rivero, Assistant Director-General of the World Health Organization on behalf of the Directors-General of the Food and Agriculture Organization of the United Nations and of the World Health Organization. The FAO Panel had already met in preparatory sessions from 1 to 5 December 1977.

Dr. Tejada-de-Rivero stated that in a world in which millions of people are inadequately fed, chemical pesticides contribute substantially to increase the yield of foodstuffs.  However, exposure to these substances through food and other means has created new public health problems.  Accordingly, the World Health Organization, in close collaboration with the Food and Agriculture Organization of the United Nations has been encouraging research in pesticide toxicology and has been undertaking various other activities to promote safe practices in the use of pesticides and protect occupationally exposed people and the population at large.  The tasks of the Meeting were to provide toxicological evaluations aimed at establishing acceptable daily intakes for humans, and to recommend limits for certain pesticides residues in specific foods.  Dr. Tejada-de-Rivero indicated that the recommendations of the Joint Meeting would provide guidance to countries attempting to control the residues of pesticides in food commodities and to the Codex Alimentarius Commission and its subsidiary body, the Codex Committee on Pesticide Residues when recommending maximum residue limits (MRLs).  He finally observed that evaluation of hazards of pesticides would not only assist in protecting the health of consumers but also at the same time would contribute to the protection of health of all humankind from the pollution of general environment by chemicals.

# 1. INTRODUCTION

The Joint Meeting was held in pursuance of recommendations made by previous Meetings and accepted by governing bodies of FAO and WHO that studies should be undertaken jointly by experts to evaluate possible hazards to man arising from the occurrence of residues of pesticides in foods.

The reports of previous Joint Meetings (FAO/WHO, 1965a, 1967a, 1968a, 1969a, 1970a, 1972a, 1973a, 1974a, 1975a, 1976a, 1977a) contain information on acceptable daily intakes (ADIs), residue limits and general principles of evaluation for the various pesticides considered. The supporting documents (FAHO/WHO, 1965b, 1965c, 1967b, 1968b, 1969b, 1970b, 1971b, 1972b, 1973b, 1974b, 1975b, 1976b, 1977b) contain detailed monographs on these pesticides and include comments on analytical methods.

The present Joint Meeting was convened to consider a further number of pesticides together with requests of both a general and specific nature.

During the Meeting the FAO Panel of Experts was primarily responsible for:

(a)    reviewing data on certain pesticides and their residues;

(b)    proposing pesticides residue limits and recommending methods of analysis.

The WHO Expert Group was primarily responsible for:

(a)    reviewing toxicological and related data on certain pesticides and their residues;  and

(b)    establishing, where possible, ADIs for man for those pesticides.

The Joint Meeting also evaluated theoretical potential daily intakes of the pesticides in relation to their ADIs and made a number of general recommendations, some of which were designed to indicate, stimulate  and coordinate lines of research.

## 2. GENERAL CONSIDERATIONS

### 2.1 MODIFICATION OF THE AGENDA

The agenda was extended to permit consideration of bromophos-ethyl, carbaryl, dimethoate, dodine, heptachlor, malathion, propoxur and sec-butylamine in relation to residue data.

### 2.2 GENERAL PRINCIPLES FOR ALLOCATING ADI FOR MAN

The Meeting accepted the principle previously established (FAO/WHO 1974a, p. 12) of expressing ADIs for man as single significant figures. All calculations from no-effect levels and safety factors for the pesticides evaluated by the present Meeting have, therefore, been rounded off to express the ADI for man in these terms.

### 2.3 QUALITY OF TOXICOLOGICAL DATA

Although most of the toxicological data available for the safety evaluation of pesticides were satisfactory, those for a number of compounds were poor or dubious quality. Other data were derived several years ago and, consequently, did not comply with the more precise standards and practices now expected. As a consequence of this many difficulties arose in the setting of new ADIs and in the re-evaluation of existing ADIs for man.

Studies on the safety evaluation of pesticides should comply with currently accepted scientific standards and with acceptable laboratory practice. Although all data submitted concerning safety evaluation will, in normal circumstances, be considered valid, occasions may arise when submitted data may be unacceptable because of low scientific standards or undesirable laboratory practices. Studies from laboratories where such unacceptable standards and practices are known to exist, or to have existed for a definite period would still be considered but might require further validation before acceptance.

Studies, conducted in more than one laboratory organization, which support or confirm each other, are likely to have more credibility than uncorroborated data from a single laboratory. Likewise, published data, which provides sufficient information for adequate evaluation and which has been subjected to review, is likely to be more readily accepted than unpublished work.

## 2.4  SAFETY FACTORS

The Meeting wished to clarify the situation regarding safety factors in arriving at
ADIs for man.  The establishment of ADIs for man is not a simple arithmetical exercise
based on the no-effect level, as the safety factor may vary widely from one compound to
another.  Although safety factors are determined empirically, they are dependent on the
nature of the compound, the amount, nature and quality of the toxicological data available, the nature of the toxic effects of the compound, whether an ADI or temporary ADI
for man is established, and the nature of any further data required.

## 2.5  COMPOSITION OF TECHNICAL PESTICIDES

The Joint Meeting has always placed great importance on the proper understanding of
the composition of technical pesticides used in the formulation of commercial products
and has always judged that well-defined technical grades should be used when conducting
toxicological and residue studies.

Information on the nature and level of impurities, intermediates and byproducts in
technical pesticides has generally been provided by manufacturers who have often indicated
that such information is regarded as a "trade secret" and of considerable value to com-
petitors.  The Meeting confirmed that it was essential for it to have access to such
detailed information but agreed that it was not usually necessary for the information to
be published in its Report and Monographs.  The Meeting was pleased to receive information
about the composition of numerous pesticides on this basis.

## 2.6  AVAILABILITY OF TOXICOLOGICAL AND RESIDUE DATA

Once again the Meeting was confronted with the difficulty of making recommendations for
maximum residue limits for many important food commodities grown extensively in tropical
regions and in developing countries.  It was recognized that the food industry, including
plantations and processors, often had information that would be valuable for the development of recommendations for maximum residue limits and every effort and support should
be given to enable the data to be made available.

The absence from Meetings of information and data known to exist and to be relevant
to the safety evaluation of pesticides resulted in the postponment of the setting of ADIs
for man for a number of compounds under consideration by the Meeting.  This lack of information would sometimes have been due to the reluctance of a manufacturer to provide data and
sometimes to the inability of FAO and WHO to seek out and obtain all the information required.
In future, data-generating sources should be invited to present comprehensive summaries
together with the detailed data submitted and also to conduct literature searches for
additional relevant information concerning the pesticides under consideration.  Members
invited to future meetings are also requested to draw attention to any additional information that is available.

## 2.7  CONSOLIDATION OF RECOMMENDATIONS

Acting on the recommendations of the 1976 meeting (FAO/WHO 1977a, p. 18) the Meeting
examined a draft of a consolidated table containing all current evaluations and recommendations of Joint Meetings, including those from the current Meeting.  It was recommended
that this compendium should be published at an early date as a separate document.  The
Meeting was advised that the Codex Secretariat was developing a system whereby the recommendations of the Codex Committee on Pesticide Residues  would be issued in a form that could

be updated regularly as new recommendations were added and other amendments made. It was suggested that the possibility of combining these operations of the CCPR and JMPR should be examined. There was general agreement that lists showing the maximum residue limits for all pesticides on each commodity and the ADIs for man would be most valuable and it was hoped that such lists could be elaborated.

## 2.8 DESCRIPTIONS OF FOOD COMMODITIES

The Meeting took note of the section of the Report of the 9th Session of the Codex Committee on Pesticide Residues, 1977 (ALINORM 78/24 paras 15-25) dealing with the ongoing efforts to establish classifications which could lead to the development of:

(a)   a uniform nomenclature for individual foods;

(b)   definitions of food groups;

(c)   definitions of the parts of the food to which the maximum residue limit applied.

In the light of the consideration given to this matter in 1976 (FAO/WHO 1977a, para 2.5) the Meeting attempted as far as was possible to adopt the names used to describe food commodities in the Codex document. The Meeting indicated that it was working towards uniformity in terminology so as to be consistent in the terms used from year to year, but recognized the need to review what has been done in the past with a view to eliminating inconsistencies.

Satisfaction was expressed at the move towards the development of concepts of grouping commodities for which a common maximum residue limit would apply and it was agreed that the Joint Meeting should work towards the establishment of recommendations for group tolerances as far as it was possible on the basis of available data.

## 2.9 MAXIMUM RESIDUE LIMITS (MRLs) IN PROCESSED FOODS

The Meeting again considered the question of maximum residue limits in processed foods (FAO/WHO 1977a, para 2.6) and confirmed that, for the purpose of establishing and enforcing maximum residue limits, raw agricultural commodities include, among other things, fresh fruits, whether or not they have been washed, waxed or otherwise treated in their unpeeled or natural form; vegetables in their raw or natural state, whether or not they have been stripped of their outer leaves, washed, waxed or otherwise treated in their unpeeled form grains, nuts, eggs, raw milk, meats and similar agricultural produce. Whilst under this definition raw agricultural commodities do not include foods that have been processed, fabricated or manufactured, e.g., by cooking, freezing, dehydrating or milling, the Meeting confirmed that it should continue to recommend maximum residue limits also for some partly processed commodities such as milled cereal products and vegetable and animal fats, which are important items of international trade, together with certain animal feedstuffs which could contain pesticide residues.

In the case of other processed foods for which no specific residue limits have been recommended the concentration of the pesticide residue on the dried, processed, cooked or preserved food when ready to eat should not be greater than the maximum residue permitted in the equivalent weight of the raw agricultural commodity. It is recognized that processing and cooking generally removes or destroys a substantial mount of the residue present on the raw commodity.

Where it is known that the residue is preferentially held in certain portions of the raw commodity, e.g., the oil, bran, peel, etc., the Meeting will continue to make recommendations by which it is possible to judge the effect of processing and utilization of wastes or specific fractions on the level and fate of residues in food.

The Meeting discussed the need for a clear description of the form or state of the commodity to which the maximum residue level applies in order to guide food control officials in preparing samples for analysis, e.g., in the case of bananas the sample should not include crown tissue or stalk but should include the skin unless otherwise indicated. In stressing the importance of guidelines on this matter, the Meeting recognized that the Codex Committee on Pesticide Residues had recently undertaken the preparation of guidelines on residue trials, on sampling and on analysis. The hope was therefore expressed that these guidelines would include appropriate information by which the sample could be defined.

## 2.10 STAGE AT WHICH MAXIMUM RESIDUE LIMITS (MRLs) APPLY

The Codex Committee on Pesticide Residues invited the Joint Meeting to define the stage at which the MRL applies (ALINORM 78/24, para 11).

In the report of the 1972 meeting (FAO/WHO, 1973a, page 41) it states that "unless otherwise indicated, the maximum residue limits should apply as soon as practicable after harvest to the raw agricultural products moving in commerce and prior to processing. For commodities entering international trade, maximum residue limits are applicable, unless otherwise indicated, at the point of entry into a country or as soon as practicable thereafter."

Generally, the data on which the Meeting determines appropriate maximum residue limits are from samples taken at harvest following good agricultural practice and the approved interval between last application and harvest. The recommendations thus refer to the amount of residue that should not be exceeded on raw agricultural produce entering trade channels immediately after harvest. Where information is available to indicate the anticipated decline in residues in storage and transport this will be indicated in the monographs but it would be inappropriate to take this into consideration in setting legal limits because of the wide range of conditions that operate in trade.

It was recommended that this concept should be expressed when the glossary (FAO/WHO, 1976a, Annex 3) is next revised.

# 3. SPECIFIC PROBLEMS

## 3.1 ETIOLOGY AND PATHOGENESIS OF LIVER TUMOURS

In considering the re-evaluation of aldrin, dieldrin and chlordane, the formation of hepatomas by these compounds in certain strains of mice was fully considered. These liver tumours have not been found to develop in any species other than mouse as the result of exposure to dieldrin or chlordane, and these chemicals have been shown to be non-mutagenic in a variety of studies.

The etiology and pathogenesis of liver tumours in mice have been discussed by the Meeting on several occasions. This Meeting re-affirmed the previous views expressed that in many strains of mice such tumours develop spontaneously, at relatively high incidence and without intentional exposure to chemicals. Moreover, the liver tumours arising in mice as the result of exposure to dieldrin or chlordane, or to other chemicals such as phenobarbitone, are very similar in nature to the spontaneous tumours (WHO/FAO, 1971a, p. 8).

The characteristic biological effects of chemicals which produce hepatomas in mice are the induction of the liver microsomal enzymes, proliferation of the endoplasmic reticulum and liver hypertrophy, all of which occur generally only at high dosage. These biochemical and cytological changes occur in other animals treated with these chemicals but liver tumours do not develop (WHO/FAO, 1974a, pp. 13-14).

In the rat, the formation of liver tumours following exposure to a variety of chemicals would seem to correlate with degranulation of the endoplasmic reticulum of the liver cells and with subsequent changes in the microsomal enzymes (Parke and Gray, 1978 a). Degranulation of endoplasmic reticulum is known to be associated with carcinogenesis (Parke, 1977 b; Parke and Symons, 1977 c; Parke, 1978 d). Both dieldrin and phenobarbitone degranulate the hepatic endoplasmic reticulum of $CF_1$ mice, a susceptible strain to dieldrin-induced tumorigenesis, but do not degranulate the endoplasmic reticulum of LACG mice, a non-susceptible strain, nor that of rat or man.

---

a   Parke, D.V., and T.J.B. Gray — A comparative study of the enzymic and morphological changes of livers of rats fed butylated hydroxytoouene, safrole, Ponceau MX or 2-acetamidofluorene. Proceedings of the 25th Symposium of Primary Liver Tumours, Titisee, October 1977

b   Parke, D.V. — Biochemical aspects. In:Principles of Surgical Oncology. Ed. R.W. Raven, Plenum Medical Press, New York and London, pp. 113-156, 1977

c   Parke, D.V. and A.M. Symons — The biochemical pharmacology of mucus. In:Mucus in Health and Diseases, pp. 423-441. Eds. M. Elstein and D.V. Parke, Plenum Medical Press, New York and London, 1977

d   Parke, D.V. — The role of the endoplasmic reticulum in carcinogenesis. Ecotoxicology and Environmental Quality. Academic Press, New York, N.Y., 1978 (in press)

A wide variety of halogenated organic compounds are known to produce this type of liver tumour in mice, which may be attributed to one or more of the following factors:

(a)    the more rapid rate of metabolic activation of these compounds in the mouse;

(b)    potentiation of the tumorigenicity of the known high level of endogenous oncogenic viruses in this species;

(c)    genetic differences in the pathways of metabolism;  or

(d)    genetic differences in the stability of the rough endoplastic reticulum.

Metabolic dehalogenation of these compounds may give rise to reactive intermediates which alter the enzymes of the endoplasmic reticulum, causing enzyme induction and membrane hypertrophy in most species, but possibly resulting in hepatic tumours in mice.

It would be unwise to classify a substance as a carcinogen solely on the basis of a species or strain-specific increased incidence of tumours of a kind that occur spontaneously with high frequency.

Furthermore, it was suggested that where man might be exposed to a chemical which resulted in an increased incidence of liver tumours in mice, carcinogenecity tests in two species other than the mouse would be appropriate (WHO/FAO, 1974a, p. 14).  Such studies have been carried out in the rat and hamster and these have shown no evidence of increased tumorigenicity.  The Meeting, therefore, felt that the production of hepatic tumours by dieldrin and chlordane in the mouse was a species-related phenomenon and were therefore able to confirm the ADI for man for these pesticides.

## 3.2 PESTICIDE RESIDUES IN RELATION TO THE TOTAL BODY BURDEN OF HALOGENATED ORGANIC CHEMICALS

Concern was expressed by the Meeting at the total environmental and human body burden of halogenated organic chemicals, and their possible chemical and biological interactions.  These substances at relatively high dosage are known to produce hypertrophy of the liver cells, proliferation of the endoplasmic reticulum and induction of the hepatic microsomal enzymes in many mammalian species.  Although in certain circumstances single halogenated organic chemicals may or may not present a hazard to man, in combination and in the presence of other chemicals they could present a greater hazard because of the overall dosage or chemical or biological interactions.

It has already been recommended that the significance and interrelationship of these biological phenomena should be investigated with reference to the health hazards of certain pesticides for man (WHO/FAO, 1971a, p. 20 and 1974a, p. 25).  The Meeting, therefore, proposed that the WHO should convene a special meeting to consider this phenomenon and the associated potential hazards to human health.

## 3.3 CARCINOGENICITY STUDIES

The concern of past meetings with the possible carcinogenic and mutagenic potential of pesticides (FAO/WHO, 1974a, 1976a) was noted, and the Meeting advised that evaluation of carcinogenic potential should be undertaken for those compounds which leave substantial residues on crops directly or indirectly used for human food.  Evaluation of carcinogenic potential should also be undertaken on those pesticides which:

(a)   have a chemical structure similar to known carcinogens, or which give rise to metabolites or residues which are known carcinogens, or closely related compounds;

(b)   give rise to pathology which is suggestive of potential tumorigenicity;  or

(c)   have pharmacokinetic properties suggestive of covalent binding to tissue or bioaccumulation.


Further information concerning the correlation between carcinogenicity and mutagenicity *in vitro* and *in vivo* is required.  Where a pesticide is shown to be a mutagen, carcinogenicity studies should be undertaken.


## 3.4   DITHIOCARBAMATE FUNGICIDES

The dimethyldithiocarbamates ferbam and ziram, and the tetramethylthiuram disulphide thiram, have similar metabolic and toxicological properties and, therefore, have been considered together under "Dimethyldithiocarbamates".  The ethylenebis-dithiocarbamates, e.g., maneb, mancozeb and zineb, are all degraded or metabolized to ethylenethiourea and have similar toxicity:  they are therefore considered together under "Ethylenebisdithiocarbamates".  Nabam, the disodium ethylenebisdithiocarbamate is also degraded to ethylenethiourea but is much more acutely toxic than other members of this class;  it has, therefore, been considered separately.  Propineb, a propylene-bisdithiocarbamate, degrades to propylenethiourea and has also been considered separately.


The Meeting recognized the importance and extent of use of this group of compounds including the fact that a number of the group had been in use for many years and that several had other industrial applications.  In view of the overall importance of making a comprehensive evaluation of the whole group as well as each individual compound and the knowledge that there was a wealth of information available in industrial archives as well as in scientific literature and government agencies the Meeting proposed that the "Groupement International des Associations Nationales de Fabricants de Pesticides" (GIFAP) be asked to establish a working group to collect and collate data from all possible sources, especially those of chemical industry (See also "Dithiocarbamate Fungicides", in Section 4.25).


## 3.5   BRAN FROM WHEAT AND OTHER CEREALS

In recent years it has been recognized that unacceptability high losses of stored grain and other non-perishable commodities due to ingestion by stored product insect pests could be minimized by the judicious use of grain protectant insecticides.  Although pyrethrum was used in limited amounts, the introduction of malathion in about 1960 revolutionized grain storage practice in many countries.

The development of malathion-resistant strains of some stored-product pests and the
lack of effectiveness of malathion against other species has prompted the search for
other suitable and acceptable insecticides. Bioresmethrin, bromophos, carbaryl, chlorpyrifos-methyl, fenitrothion and pirimiphos-methyl have been evaluated for this purpose
and these insecticides have been adopted in one or more countries for the protection
of raw grains in storage.

Extensive studies have shown that the insecticide deposits on the grain do not
penetrate far beyond the seed coat and, therefore, much of the deposit is removed in
cleansing and milling. In the case of rice in husk the bulk is removed with the husk
and remainder in the bran. In the case of wheat more than 75 percent of the deposit
on the grain at the time of milling is usually removed in the bran. The concentration of
the residue in the bran may be two - four times the concentration in the whole grain and
the Meeting has recommended maximum residue limits which reflect this fact.

The Codex Committee on Pesticide Residues at its 9th Session (1977) asked for
clarification of doubts concerning the apparently high residues in bran (ALINORM 78/24,
para 154). In the 1974, 1975 and 1976 monographs (FAO/WHO, 1975b, 1976b, 1977b) there
are numerous studies showing the distribution and fate of grain protectant insecticide
residues in bran and it is apparent that in commercial practice crude bran of the type
used in limited quantities for dietetic purposes will generally contain not more than
5 mg/kg of insecticide residues (even though some will contain residues up to the
recommended MRL). In the preparation of flour for industrial purposes (bread and
biscuit making, cake flour, pasta production) wheat from several sources is usually
blended in order to obtain the qualities desired in the flour and baked goods. Such
blending will usually lead to a dilution of the residues, especially those segregated in
bran.

Information available to the Meeting suggests that the total intake of bran including dietetic use is normally not greater than 25 g/person/day. A consumption of 50 g/d
would be exceptional. In view of the large quantity of bran available and the small
amount needed for processing and distribution for dietetic use manufacturers might be
encouraged to select batches which have below-average levels of insecticide residues.

There appears to be no significant consumption of rice bran other than as animal
feed.

## 3.6 NITROSAMINES IN PESTICIDE FORMULATIONS

The Meeting was aware that the presence of nitrosamines recently have been reported
in a number of pesticide formulations a. Full review was deferred until the problem is
more completely defined. Sufficient information was available to indicate that the presence of nitrosamines can be diminished if nitrite corrosion inhibitors are not added
to the formulations. Additional information indicates that the nitrosamine content can
be minimized by modification of manufacturing processes, particularly with respect to
the sequence of nitration and amination steps.

---

a    Fine, D.H., et al - Determination of N-Nitroso Pesticides in Air, Water and Soil.
Proceedings of the 172nd American Chemical Society National Meeting, San Francisco,
California, 2 September, 1976

## 4. EVALUATION OF DATA FOR ACCEPTABLE DAILY INTAKE
## FOR HUMANS AND MAXIMUM RESIDUE LIMITS

EXPLANATION

This section contains notes on the compounds considered by the present Meeting.  The information is more detailed than in previous reports, particularly in the case of compounds evaluated for the first time, and provides a summary of the material which will subsequently appear in the full monographs.  (Evaluation of Pesticide Residues in Food, FAO/WHO, 1978).

As in the monographs on the compounds considered at previous Meetings, compounds which were re-evaluated by the present Meeting are marked with an asterisk.

Compounds evaluated for the first time are identified by their chemical names, according to IUPAC nomenclature, as well as their common names.

The recommended maximum acceptable daily intake for man (ADI) and maximum residue limits (MRLs) are tabulated in Annex I, Part I.  Guideline Levels for compounds for which no ADI has been allocated appear in Annex I, Part II.

Details of further work or information considered necessary or desirable by the Meeting are listed separately in Annex 2.  The requirements replace those mentioned in earlier reports unless otherwise stated.

## 4.1 ALDRIN/DIELDRIN

Toxicology

The required teratogenicity studies in different animal species were submitted and showed dieldrin not to be a teratogen.  In several studies, where the test compounds were administered in repeated small doses to pregnant mice no teratogenic effect could be found.  However, the administration of a single high dose of 15 mg/kg  dieldrin to mice ($1/2$ $LD_{50}$) or 30 mg/kg to hamsters resulted in minor malformations which were attributed to maternal toxicity.

Results of various carcinogenicity tests in mice and other mammalian species indicate that there is a species-specific effect of dieldrin and aldrin on the mouse liver, resulting in an increased frequency of liver tumours only in this animal species.  In all species examined (mouse, rat, rabbit, rhesus monkey, chimpanzee) 12-hydroxy-dieldrin and 4,5-aldrin-trans-dihydrodiol were the major metabolites.  Regarding the ratios of these two metabolites the mouse seems to metabolize the test compound predominantly by the opening of the epoxide ring.  This metabolic pathway on the one hand and a high rate of metabolism compared to other animal species examined on the other, could result in relatively high concentrations of 4,5-aldrin-transdihydrodiol in the mouse liver.

Dieldrin, as well as other chlorinated pesticides, is a powerful inducer of micro-somal enzymes. Owing to their high rate of metabolism, mice are especially susceptible to such effects. (See Section 3.1). Compared to other animal species, it reacts rather anomalously and the mouse, therefore, may not be an appropriate model for man in this case.

The results of prolonged enzyme induction are proliferation of the endoplasmic reticulum in the cells, and hypertrophy and hyperplasia of the liver. These changes are fully reversible if treatment is ceased before tumours have developed on the chronically overloaded and damaged cells. Similar effects are produced by phenobarbital.

The results of the different experiments for the study of the possible mechanism of the tumorigenic action of dieldrin on mouse liver at the subcellular level are inconsistent.

A number of in vitro and in vivo mutagenicity tests have been performed. In none of these studies did dieldrin reveal any mutagenic activity.

These new findings again support the view that dieldrin and aldrin are not carcino-gens on the basis of knowledge available to the Meeting. Therefore, no change of the existing ADI for man was considered necessary.

TOXICOLOGICAL EVALUATION

Level causing no toxicological effect

   Rat:   0.5 mg/kg in the diet, equivalent to 0.025 mg/kg bw
   Dog:    1 mg/kg in the diet, equivalent to 0.025 mg/kg bw

Estimate of Acceptable Daily Intake for Man

   0-0.0001 mg/kg bw 1/

4.2 AMITROLE *

Toxicology

Several new studies on mutagenicity became available. No mutagenic action could be demonstrated. In a recent carcinogenicity study with a very high dose level, which had an influence on survival, a carcinogenic action on the thyroid was found. In addition, two cases of cholangio-fibrosis in the liver have been observed. Two metabolites of amitrole have been identified after oral administration to rats.

---

1/ The ADI is applicable to aldrin and dieldrin separately or to the sum of them
   if both are involved

In four short-term studies the main effect was an increased uptake of radioactive iodine by the thyroid at dose levels of 50, 200 and 500 ppm. This uptake correlated well with the histologically observed activation. With 20 ppm only a very marginal effect was found while 2 ppm was without effect.

This Meeting confirmed the existing conditional ADI for man pending consideration of the whole concept of conditional ADI for man at a future meeting.

## 4.3 BROMOPHOS *

### Residue and Analytical Aspects

At the 9th Session of the Codex Committee on Pesticide Residues, the Joint Meeting was requested to review the proposed MRLs for red currants in the light of the proposed figures for blackberries and black currants, and to propose a MRL for meat in the light of the use of bromophos on sugar beet.

In response to the request of the CCPR, the data considered by the 1972 meeting were reviewed. They indicate that residues on black currants, blackberries and red currants 7 days after application of the same dosages of bromophos show relatively wide variations. In the light of this, the MRLs for blackberries and black currants were raised to 1 mg/kg.

No data were available on which to judge the likelihood of transfer of bromophos residues to meat from the feeding of leaves of sugar beets.

### Toxicology

Bromophos was previously evaluated and a temporary ADI for man of 0–0.006 mg/kg bw was allocated.

An adequate study in dogs previously requested has been provided. Renal function tests and testicular histology did not confirm previous concern. Plasma cholinesterase was more susceptible to depression than erythrocyte cholinesterase; however, the previous study revealed the opposite. A no-effect level was demonstrated in the 20 ppm group.

Results of a previously requested carcinogenic study have also become available. In this long-term carcinogenicity study in mice, no compound-related effect was observed.

A 28-day study in human volunteers showed no significant effect of bromophos at a level of 0.4 mg/kg bw/day based on plasma cholinesterase activity.

Attention was paid to the 4-bromo-2,5-dichlorophenol part of the bromophos molecule. The available information indicated that the synthesis of this phenolic compound does not incur any risk of forming p-dibenzodioxins.

As the required studies were provided and observations in man were available, a higher ADI for man was allocated.

TOXICOLOGICAL EVALUATION

Level causing no toxicological effects

    Dog:    20 mg/kg in diet, equivalent to 0.5 mg/kg bw

    Man:    0.4 (mg/kg bw)/day

Estimate of Acceptable Daily Intake for Man

    0-0.04 mg/kg bw

4.4  BROMOPHOS-ETHYL *

Residue and Analytical Aspects

The 9th Session of the Codex Committee on Pesticide Residues (1977) increased the MRLs on black currants, strawberries, cabbage and Kohlrabi to 0.5 mg/kg. The Meeting re-evaluated the information on these commodities and increased the limits similarly. As requested by Codex the Meeting considered the effect of these changes and judged that it was negligible.

4.5  sec-BUTYLAMINE *

Residue and Analytical Aspects

Following evaluation of this compound in 1975, the Joint Meeting considered that additional information was required before the temporary MRLs could be confirmed. Some of this information was available to the Meeting.

In spite of early indications that sec-butylamine was effective against a range of fungal organisms which are detrimental to a number of fruits and vegetables, it is apparently not being utilized commercially on fruits other than citrus on which there was a complete monograph in 1975. There has been limited use on potatoes for the control of gangrene and skin spot. Although it is known that significant residues remain, no detailed information was available. There is no indication that these residues transfer into the tubers that develop from seed potatoes that have been treated; but it would be desirable to have information on the levels and fate of residues on edible potatoes treated before storage.

Cows receiving dried citrus pulp containing normal commercial residues or excessive dosage of sec-butylamine excreted most of the residue unchanged in the urine but milk contained from 0.071 to 0.67 mg/kg of sec-butylamine (mean 0.33 mg/kg). There was no tendency for residues to accumulate as a result of repeated administration. Traces of sec-butylamine could be detected in muscle, blood and fat of cows slaughtered immediately at the end of a period of continuous ingestion of sec-butylamine residues. Significant residues (1-2.7 mg/kg) were found in kidney and lesser amounts (0.15-0.2 mg/kg) in liver.

sec-butylamine occurs naturally at concentrations up to 0.3 mg/kg in fresh milk and much higher in sour milk together with significant quantities of other engodenous amines. Extensive isolation and purification procedures are required to separate, recover and measure residues of sec-butylamine resulting from the ingestion of feeds containing residues of the fungicide.

It was considered that the information received fulfilled many of the requirements listed in 1975 and pending the full re-evaluation scheduled for 1978, temporary MRLs were proposed to cover the residues resulting from the feeding of citrus pulp and citrus molasses.

## 4.6 CAPTAFOL *

### Residue and Analytical Aspects

The Meeting was requested by the Codex Committee on Pesticide Residues at its 9th Session (1977) to re-examine the data on which recommendations of the 1973 Joint Meeting for maximum residue limits on cranberries, apples and pears were based. The Meeting re-affirmed the previous recommendation of 8 mg/kg on cranberries based on a 50-day pre-harvest interval. The current MRL of 5 mg/kg on apples and pears was retained until additional information on national use patterns and residue trials becomes available.

### Toxicology

Captafol appears to be not mutagenic in a dominant lethal test in mice and host-mediated assay in rats.

In a two-year study with rats a non-dose related increase in the relative liver weight was evident in males after 12 months, but not at 24. The absolute liver weight of the males at 12 months was also significantly increased at 250 ppm but this effect was not produced in the 500 and 1 500 ppm groups. The histopathological changes observed in the liver and kidneys at 1 500 and 5 000 ppm were considered to be features of long-term toxicity, sometimes seen in old rats.

It seems justified, therefore, to consider 250 ppm as a no-effect level in rats. The temporary ADI for man has been changed to an ADI for man at a higher value.

### TOXICOLOGICAL EVALUATION

### Level causing no toxicological effect

Rat: 250 mg/kg in the diet, equivalent to 12.5 mg/kg bw

Dog: 10 (mg/kg bw)/day

### Estimate of Acceptable Daily Intake

0-0.1 mg/kg bw

## 4.7 CAPTAN *

### Residue and Analytical Aspects

In response to a request from the 9th Session of the Codex Committee on Pesticide Residues (1977), the previously recommended MRL of 40 mg/kg on apples, pears and cherries was re-examined. It was concluded that the MRL on apples and pears should be reduced to 25 mg/kg, but the present MRL on cherries should be maintained until additional information on national use patterns and supervised trials is submitted.

<u>Toxicology</u>

The Meeting was aware of the existence of recent studies on the mutagenicity and carcinogenicity of this compound.  Since the additional information was not available in detail, consideration of this pesticide was deferred to a future meeting.

## 4.8  CARBARYL *

<u>Residue and Analytical Aspects</u>

Following the review in 1976 when maximum residue limits were recommended for a number of raw grains, further data on the level and fate in stored sorghum, and information on the fate on stored grain generally, provided a basis for judging that the maximum residue limit recommended previously for carbaryl on sorghum from pre-harvest use was adequate to cover residues from post-harvest application.

## 4.9  CARBENDAZIM *

<u>Toxicology</u>

Carbendazim is formed during the degradation of benomyl and thiophanate methyl to which an ADI for man was previously allocated.

Malignant lymphomas were induced after 20 - 43 weeks in mice whose mothers were treated with carbendazim combined with nitrite.  Treatment with carbendazim alone was ineffective.

No indication of mutagenic activity was found in a dominant lethal test and in cytogenetic analyses of the bone marrow of rats treated with up to 1 000 ppm of carbendazim in the diet.  In a micro-nucleus test, however, a dose-related increased number of nucleated erythrocytes was observed.

Since the required toxicological data were not submitted, no recommendation for an ADI for man can be made, and decisions on this compound was postponed to a future meeting.

## 4.10 CARBOPHENOTHION *

<u>Residue and Analytical Aspects</u>

Since the present Meeting allocated a temporary ADI for man (see below), the previously recorded guideline levels were converted to temporary MRLs.

<u>Toxicology</u>

Carbophenothion was previously reviewed and further studies to substantiate the marked species difference in sensitivity to plasma cholinesterase depression, and an adequate reproduction study were required.  In 1976 the temporary ADI for man was

withdrawn because the information previously requested had not been provided.  Part
of this information has now been received and has been considered together with re-
evaluation of short-term and long-term rat and dog studies in which cholinesterase
activity was determined.

Present carbophenothion metabolism studies confirm and extend the studies previously reported.  Results show that carbophenothion is readily degraded both in the
rat and the goat, primarily to water soluble products, which are excreted in the urine.
None of the oxygen analogs of carbophenothion were detected.  The acute toxicity of the
major metabolites of carbophenothion was considerably lower than that of carbophenothion.

From both the data previously reported and the present data it is clear that plasma
cholinesterase inhibition is the most sensitive criterion in both short- and long-term
studies in rats and dogs.  In a two-year study in dogs 5 ppm or 0.125 mg/kg bw/day was
not a no-effect level with respect to plasma cholinesterase inhibition.  In a 90 day
experiment in dogs an effect was found even at 0.04 mg/kg bw/day, while 0.02 mg/kg bw
was a marginal no-effect level in this respect.  In a two-year experiment with rats
5 ppm in the diet (0.25 mg/kg bw/day) caused an inhibition of RCB-cholinesterase after
13 and 26 weeks.

A three generation reproduction study including a teratological study revealed
a no-effect of 3 ppm equivalent to 0.15 mg/kg bw.  With respect to the lowest no-effect
level this study has no consequence for the calculation of the acceptable daily intake
for man, the most sensitive criterion still remaining is the plasma cholinesterase
depression in dogs.  An extreme difference in sensitivity between man and dog is noticed
in the reported cholinesterase depression studies.  There seems to be an indication
that 0.8 mg/kg/day for 30 days did not result in cholinesterase inhibition in man.
However, 1 mg/kg bw in a calf caused symptoms and severe cholinesterase inhibition.
The Meeting decided to allocate a temporary ADI for man at a lower value owing to the
marked species difference in sensitivity to plasma cholinesterase depression.

TOXICOLOGICAL EVALUATION

Level causing no toxicological effect

    Rat:  3 mg/kg in the diet, equivalent to 0.15 mg/kg bw

    Dog:  0.02 (mg/kg bw)/day

Estimate of Temporary Acceptable Daily Intake for Man

    0.0002 mg/kg bw

## 4.11 CHINOMETHIONAT *

Residue and Analytical Aspects

The compound was re-evaluated in the light of further information previously requested and the request by the 9th Session of the Codex Committee on Pesticide Residues
for a lowering of the temporary MRL in apples.  A review of the data submitted showed
that the temporary MRL can be lowered from 0.5 mg/kg to 0.2 mg/kg.  Most of the residues
in apples and oranges are confined to the peel.

Residues in the meat, milk, liver and kidneys of cows fed in the diet with dry citrus pulp or almond hulls containing chinomethionat were at or below the limit of determination.

The manufacturer submitted a colorimetric procedure for the determination of free 2,3-dithiolo-6-methylquinoxaline (QDSH) at concentrations above 0.1 mg/kg. The lower limit of determination using the Vogeler method for the parent compound is 0.02 mg/kg in cucumbers, berry and pome fruits.

## Toxicology

A temporary ADI for man was previously established.

A request was made for further metabolic studies in a non-rodent species, explanation of the liver cell hyperplasia, loss of liver microsomal enzyme activity and, where possible, observations in man. Although some new metabolic data have been supplied, in general no other information was provided. Therefore, it was decided to extend the temporary ADI for man.

## TOXICOLOGICAL EVALUATION

### Level causing no toxicological effect

Rat:  12 mg/kg in the diet, equivalent to 0.6 mg/kg bw

### Estimate of Temporary Acceptable Daily Intake for Man

0–0.003 mg/kg bw

## 4.12 CHLORDANE *

### Residue and Analytical Aspects

The 1977 Codex Committee on Pesticide Residues asked for a reduction of the proposed MRLs for the items contained in ALINORM 78/24, 12.16–12.31. No data being available to the 1977 Joint Meeting to justify such a reduction, the existing recommendations were not changed.

The Meeting considered that there was no reason to consider the proposed MRLs as extraneous residue limits as they can result from registered uses.

### Toxicology

Chlordane is a member of the group of organochlorine insecticides which have the common effect of inducing liver microsomal enzymes (see Section 3.1 and 3.2).

In long-term studies, chlordane caused hepatocellular carcinoma in mice at a dose of 60 ppm in the diet, but not in rats at doses as high as 400 ppm for males and 240 ppm for females. Both male and female rats were found to have an increase in proliferative lesions of follicular cells of the thyroid. Male rats only showed an increase of malignant fibrous histiocytomas, but these observations were discounted because of their low incidence and the high variability of the number of tumours in earlier experiments at the same research centre.

In a short-term test with liver enzyme induction as the most sensitive criterion, a no-effect level in rats was established which is lower than the one of 20 ppm earlier established.


TOXICOLOGICAL EVALUATION

Level causing no toxicological effect

    Rat:  5 mg/kg in the diet, equivalent to 0.25 mg/kg bw

    Dog:  3 mg/kg in the diet, equivalent to 0.075 mg/kg bw

Estimate of Acceptable Daily Intake for Man

    0-0.001 mg/kg bw


## 4.13 CHLORDIMEFORM *

Toxicology

It was brought to the attention of the 1976 Meeting that the manufacturers of chlordimeform have voluntarily suspended production and marketing of the product for the time being. The Meeting was also informed of the probable effect of 4-chloro-o-toluidine, a major metabolite of chlorodimeform, in inducing haemangiosarcoma in mice. In addition, the 1975 Meeting expressed concern about a preliminary report on haemorrhagic cystitis in occupationally exposed people. No new data have been received which alleviate the concerns of these Meetings. It was, therefore, not possible to change the temporary ADI of 0.01 and this Meeting decided to take no action concerning this compound.

Further toxicological evaluation has been postponed until a subsequent Joint Meeting has received relevant information addressing the concerns of the 1975 and 1976 Meetings.


## 4.14 CHLOROBENZILATE *

Residue and Analytical Aspects

In response to the need for further information mentioned by the 1975 Joint Meeting, additional data were provided.

When chlorobenzilate was applied to a bare, silty loam soil at 5 kg a.i./ha, disappearance was rapid (t1/2 <30 days) and vertical movement was confined to the upper 5 cm of soil. The metabolites 4,4'-dichlorobenzilic acid and 4,4'-dichlorobenzophenone reached maxima after 20 and 61 days respectively, and rapidly decreased thereafter.

No residues could be detected (<0.05 mg/kg) in wines made from white and red grapes containing residues near the accepted maximum residue limit for grapes.

## 4.15 CHLOROTHALONIL*

### Residue and Analytical Aspects

The compound was re-evaluated in the light of information requested by the 1974 Joint Meeting and the decision of the 1977 Codex Committee on Pesticide Residues to change the temporary MRL in oranges to encompass all citrus fruits. The 1974 Meeting recommended a maximum residue limit of 30 mg/kg in peaches, but the principal manufacturer suggested that this could be reduced to 25 mg/kg. Supporting data, mainly from the United States, justifies such a reduction. In bananas, the fungicide remains mainly in the peel.

When dairy cows were fed diets containing very high levels of chlorothalonil or 2,5,6-Trichloro-4-hydroxy-1,3-benzocarbonitrile (DAC-3701), approximately 0.2% of the chlorothalonil and 23% of the DAC-3701 were eliminated in the milk as DAC-3701. Of the tissues examined, only the kidneys contained residues above the limit of determination.

After being ensiled for 18 days, maize forage was found to contain 10% of the initial residues.

Washing removed up to 94% of the chlorothalonil residues in tomatoes and up to 97% in peaches and residues could not be detected (<0.01 mg/kg) in tomato and peach processing products.

The temporary MRLs recommended in 1974 were maintained as temporary limits with the exception of that for peaches which was lowered and for "oranges" which was changed to "citrus fruit". An additional temporary limit was recommended for bananas.

### Toxicology

The data previously required on additional studies to resolve the lowest dose limits for kidney effects in the rat and the determination of growth reduction in pups relative to dietary intake or secretion into maternal milk, were submitted for consideration by the Meeting. The absence of sufficient detail in the reports on kidney pathology, and discrepancies in body weight data in the rat reproduction studies made evaluation of these data impracticable. However, since data have been submitted in response to the previous request, the Committee extended the current temporary ADI for man for an additional two years.

### TOXICOLOGICAL EVALUATION

#### Level causing no toxicological effect

Rat: 60 mg/kg in the diet, equivalent to 3.0 mg/kg bw

Dog: 120 mg/kg in the diet, equivalent to 3.0 mg/kg bw

#### Estimate of Temporary Acceptable Daily Intake for Man

0.03 mg/kg bw

## 4.16 CHLORPYRIFOS *

### Residue and Analytical Data

At the 9th (1977) Session of the Codex Committee on Pesticide Residues (ALINORM 78/24, par. 77), it was agreed that the MRL for peppers should be raised from 0.2 mg/kg to 0.5 mg/kg, since the retention of the residue on peppers was similar to that on tomatoes for which a maximum residue limit of 0.5 mg/kg had been recommended. The delegation of Israel undertook to provide data to the Joint Meeting.

Although no data were received, the Meeting concurred with the observations of the Codex Committee on Pesticide Residues and amended the recommendation accordingly.

### Toxicology

New data on biotransformation indicate that the main metabolite of chlorpyrifos in the urine is the glucoronide of 2,5,6-trichloro-2-pyridinol (80%), instead of 2,5,6-trichloro-2-pyridyl phosphate. At least four other metabolites were found. One of these was identified as the glycoside of 2,5,6-trichloro-2-pyridinol.

Chlorpyrifos showed no mutagenic activity in a metabolically activated microbiological assay system.

Although no further information has become available about the possible increased sensitivity to plasma cholinesterase depression after withdrawal from an initial dose regime, there is no base to justify a change of the ADI for man other than in accordance with comments in this report to round this off to one significant figure (see Section 2.2).

### TOXICOLOGICAL EVALUATION

### Level causing no toxicological effect

Rat:  0.03  (mg/kg bw)/day

Dog:  0.01  (mg/kg bw)/day

Man:  0.014 mg/kg bw orally for one month

### Estimate of Acceptable Daily Intake for Man

0–0.001 mg/kg bw

## 4.17 CYHEXATIN *

### Toxicology

Cyhexatin was previously evaluated and a temporary ADI for man was allocated. Further work was required to elucidate the significance of the occurrence of adenomas in rats and the diminished weight gain in animals.

It was noted that a long-term carcinogenicity study in rat was finished and that results may soon be available.  For this reason the Meeting considered that the temporary ADI for man could be extended.


## 4.18 DDT *


Toxicology

No new data were reviewed.

DDT has a conditional ADI for man.  The concept of a conditional ADI for man will be discussed at a future meeting.


## 4.19 DAMINOZIDE  (N-dimethylaminosuccinamic acid)


Residue and Analytical Aspects

The growth regulator daminozide has been used worldwide on a variety of fruits and vegetables since 1967.  Residue limits have been established in various countries.

The technical grade product of the primary manufacturer contains 98 - 99% active ingredient and is currently available only as a water soluble powder.  Daminozide was formerly manufactured by a process in which dimethylnitrosamine was a starting reactant, and occurred as an impurity in the technical grade product.  The primary manufacturer informed the meeting that a new process has been adopted in which dimethylnitrosamine is not used.  The recommended guideline limits (See Annex I, Part II) apply only to daminozide manufactured from the process in which dimethylnitrosamine is excluded.

The compound is readily absorbed and translocated from foliar sprays.  The parent molecule in free or loosely bound forms is persistent, particularly in fruits.  The chemical structure suggests the possible presence of some metabolites of interest, including nitrosamines, hydrazides, and other hydrazine derivatives.  However, no positive findings of these metabolites were reported by chemical methods of analysis. The postulated metabolite N-dimethylnitrosamine was not detected by a method said to be sensitive to 0.002 mg/kg.  Additional studies indicated indirectly the absence of unsymetrical dimethylhydrazine  (UDMH) in treated crops.

When animals were fed at levels higher than the anticipated intake from byproducts of treated crops, daminozide was found at low levels in milk, eggs, and tissues.  It was concluded that residues resulting from the maximum possible intake in a practical situation would be below the level of determination of the analytical method (0.2 mg/kg in meat and eggs, 0.05 in milk).

A colorimetric method of analysis is available which measures UDMH released by vigorous caustic hydrolysis.  The chromogenic agent is not specific for UDMH.  It is thought that the method would also measure any free UDMH and complexed daminozide.  It would not be very useful for regulatory purposes.

The extensive data available from supervised trials showed considerable variability in harvest residues, but were sufficient to support the guideline levels recorded in Annex I.

## Toxicology

Daminozide is not highly acutely toxic as measured by $LD_{50}$ values in the g/kg range for oral and dermal studies and is only mildly irritating to the eye and skin. It is rapidly excreted and does not bioaccumulate in mammals. Daminozide did not produce lethal effects in male mice at 10 000 ppm and no teratological or reproductive effects were observed at levels of 300 ppm or lower. An adequate two-year dog study has been conducted.

The long-term rat study was inadequate because of internal conflicts within the report and because of the inability to evaluate the carcinogenic potential of daminozide. Information on the biotransformation of daminozide in animals is needed.

Because of the above concerns no recommendation for an acceptable daily intake for man could be made.

## 4.20 DICHLOFLUANID *

## Residue and Analytical Aspects

Further information on dichlofluanid was requested by the 1974 Joint Meeting, especially on its fate in and on plants, on residues from supervised trials and on the suitability of Becker's analytical method for regulatory purposes.

After washing, canning or jam-making, residues in strawberries decreased by 80% or more. The Becker method has been validated for regulatory purposes. Temporary MRLs, in addition to those recommended by the 1974 meeting, were recommended.

## Toxicology

Adequate information on absorption, distribution, excretion and pharmacokinestics on the dichlorofluoromethylthio moiety has been provided. Within two days about a tenfold decrease in the amount of labelled dichlorofluanid was observed in most of the organs. After 10 days a residue of 1% of the dose was present. The consequence of relatively high residues of this organic fluorine compound in the thyroid was of some concern and should be further investigated.

TOXICOLOGICAL EVALUATION

## Level causing no toxicological effect

Rat: 1 500 mg/kg in the diet, equivalent to 75 mg/kg bw

Dog: 1 000 mg/kg in the diet, equivalent to 25 mg/kg bw

## Estimate of Temporary Acceptable Daily Intake for Man

0-0.3 mg/kg bw

## 4.21 DICHLORVOS *

### Toxicology

On the basis of recent and earlier studies, the metabolic patterns of dichlorvos in animals are clearly elucidated. It is rapidly broken down in mammals to products which are excreted or incorporated into natural biosynthetic pathways.

Several new studies have examined the potential for direct methylation of nucleic acids by dichlorvos. It has been shown to cause the methylation of bacterial and mammalian DNA and RNA _in vitro_. Also the urine from dichlorvos-treated mice and rats was found to contain methylated purines. The alkylating properties of dichlorvos led to the suspicion that dichlorvos might be mutagenic and/or carcinogenic. However, no methylated 7-methylguanine has been detected in the DNA or RNA from liver or kidney tissue or other soft tissues of rats or mice treated with $^{14}C$-methyl labelled dichlorvos.

Although dichlorvos is a potential alkylating agent of DNA and RMA _in vitro_, this potential is apparently not realized in vivo owing to the rapid degradation of dichlorvos in mammals. The presence of $^{14}C$-labelled purines in the urine of animals treated with $^{14}C$-labelled methylating agents is not conclusive proof of the methylation of DNA and RNA at the polymeric level. The $^{14}C$ may be incorporated into the free purine bases. Thus, animals treated with $^{14}C$-labelled guanine excreted $^{14}C$-labelled 7-methylguanine in the urine.

These studies and the negative results in mammalian mutagenicity and carcinogenicity studies support the view that dichlorvos has an extremely low potential for producing mutations or cancer in man. None of the recent studies have changed the basis for establishing an ADI.

### TOXICOLOGICAL EVALUATION

### Level causing no toxicological effect

Man: 0.033 (mg/kg bw)/day    Rat: 10 ppm in the diet equivalent to 0.5mg/kg/day
Dog: 0.37 mg/kg bw/day

### Estimate of Acceptable Daily Intake for Man

0-0.004 mg/kg bw

## 4.22 DICLORAN

### Residues and Analytical Aspects

Following the evaluation of dicloran in 1974, the Joint Meeting listed information that was required before MRLs could be confirmed. Some of this information has been provided and was evaluated.

The question of wide variation in residue levels in berry fruits was considered in the light of further study of the extensive data evaluated in 1974. It was judged that the variations were normal and reflected the several species of berries involved, the different cultivars of each species and the varying ecological and environmental conditions. Experience with other pesticides applied to berries has indicated that spraying technique and spray equipment can also contribute to variations in residue levels.

When exaggerated levels of dicloran  are administered to pigs and cattle, significant residues can be found in various edible tissues with the highest concentration in liver.  However, cows receiving rations containing dicloran  residues at concentrations likely to be found in agricultural commodities following the use of dicloran  produced milk without detectable residues.

Simple washing processes remove over 90% of dicloran  residues.  Cooking removes approximately 50% of the residue from beans while the canning of peaches removes all detectable residue from fruit treated both pre- and post-harvest.

No detectable residues ($<$0.01 mg/kg) of dicloran  or its metabolites were found in wine made from grapes that had been treated with dicloran   No adverse effect on fermentation or quality of wine could be observed following the treatment of grapes shortly before harvest.

Micro-organisms readily convert dicloran  into 2,6-dichloro-p-phenylenediamine and 4-amino-3,5-dichloroacetanilide.  None of several studies indicated any appreciable uptake of residues of dicloran  or its metabolites from soil by plants, including vegetables.

On a basis of the additional information before the Meeting the MRLs previously recommended were confirmed.  Additional MRLs were recommended for dicloran in onions, nectarines and chicory.

## Toxicology

Dicloran  (2,6-dichloro-4-nitroaniline) was previously evaluated and a temporary ADI for man was established.  Further studies on the oculotoxicity observed in the dog and of the fate of dicloran  in livestock animals were requested.  Further work on the effects of dicloran  on the hepatic microsomal enzymes of animals species and further observations of its effects in humans, were also considered desirable.

A study of the disposition of orally administered $^{14}C$-dicloran  in rats, dogs and pigs has shown that distribution of radioactivity resulted in substantially higher concentrations in the tissues of dog and pig than rat.  A more rapid increase in plasma concentration of $^{14}C$ in the dog than in either pig or rat, and a different spectrum of the metabolites suggests that there is a species difference in the kinetics of metabolism.  The higher concentration of $^{14}C$ in the retina and iris of the eye of the dog than occur in other dog tissues or in the pigmented tissues of the eye of rat or pig, goes some way to explaining the photosensitive oculotoxic effects of dicloran which have been observed in the dog.  It is likely that at least some of this radio-activity present in the pigmented tissues of the dog eye is attributable to the major metabolite of dicloran,  4,amino-3,5-dichlorophenol, and under the influence of light this would be oxidized to the corresponding quinoneimine which could difuse into the non-pigmented tissues of the eye, the lens and cornea, producing lesions owing to the oxidation of this metabolite.

The oculotoxic effects in the dog are, therefore, partly explained by the different kinetics of dicloran  seen in this species and the known sensitivity of the dog eye to oxidant chemicals.  No similar oculotoxic effects have been seen in rat, pig, Rhesus monkeys or man dosed with dicloran.

Dicloran, at doses of 10 mg/kg/day or more, results in induction of the microsomal mixed function oxidases of rat liver, larger doses (400 mg/kg/day) result in liver enlargement, and still larger doses (500 mg/kg/day) decreased mitochondrial oxidation, but without the uncoupling of oxidative phospherylation. Both dicloran and its major metabolite, 4-amino-3,5-dichlorophenol, inhibit mitochondrial electron transport and uncouple oxidative phosphorylation _in vitro_ but not _in vivo_, probably because of the conjugation of the metabolite(s) _in vivo_.

In mutagenicity studies in various strains of _B. subtilis_ and _S. typhimurium_ and _E. coli_, with and without activation by rat liver microsomal enzymes, dicloran was not found to be mutagenic.

In view of the further work carried out to explain the oculotoxicity of dicloran in the dog, the metabolic and pharmacokinetic studies in the pig, and the eludication of the effects of dicloran on liver microsomal enzymes, in compliance with the requests of the 1974 Joint Meeting, it was recommended that the previous temporary ADI for man be made an ADI for man at the same level.

TOXICOLOGICAL EVALUATION

## Level causing no toxicological effect

Rat:  1 000 mg/kg in the diet, equivalent to 50 mg/kg bw

Dog:  100 mg/kg in the diet, equivalent to 2.5 mg/kg bw

## Estimate of Acceptable Daily Intake for Man

0–0.03 mg/kg bw

## 4.23 DIMETHOATE *

## Residue and Analytical Aspects

Data indicating that the 1973 Joint Meeting recommendation for a residue limit of 0.3 mg/kg on strawberries was inadequate was accepted and a revised limit of 1 mg/kg was recommended.

## 4.24 DIQUAT *

## Residue and Analytical Aspects

Data are still being gathered to answer questions from the 9th Session of the Codex Committee on Pesticide Residues (1977) and it is proposed to consider them at the next Joint Meeting.

## Toxicology

Two metabolites, diquat monopyridone and diquat dipyridone, have been identified after oral and subcutaneous administration of diquat to rats. The same type of metabolites were found after i.p. injection in rabbits, guinea-pigs and rats. These metabolites are produced only in small amounts and are less toxic than diquat itself. It seems, therefore, that the parent compound is the toxicologically active substance. In a new long-term study with rats special attention was paid to the cataractogenic action of diquat. A significant increase in the number of cataracts was observed in the 75 ppm group. There was no significant difference betweeh the number of cataracts at dietary levels of 15 and 25 ppm and those on control diets. In this respect, 25 ppm of diquat ion was found to be the no-effect level. However, the cataracts, in the experimental groups, especially in the 25 ppm group appeared somewhat earlier than in the control group. Cataracts have never been observed in man following occupational exposure to diquat. These findings enabled some raising of the ADI figure.

## TOXICOLOGICAL EVALUATION

### Level causing no toxicological effect

    Rat:   15 mg/kg diquat ion in the diet (fed as dibromide), equivalent to 0.75 diquat ion/kg bw

    Dog:   68 mg/kg diquat dichloride in the diet, corresponding to 1.22 mg diquat ion/kg bw

### Estimate of Acceptable Daily Intake for Man

    0-0.008 mg/diquat ion/kg bw

## 4.25 DITHIOCARBAMATE FUNGICIDES * (Also see Section 3.4)

### Residue and Analytical Aspects

The current position regarding some chemical aspects of ethylenebis dithiocarbamate (EBDC) fungicides as a group was discussed by the FAO Panel of Experts. Particular attention was given to a recent publication (IUPAC, 1977) on ethylenethiourea (ETU) which is a manufacturing, food processing and metabolic product of the EBDC compounds. The IUPAC conclusions were fully endorsed and the recommendations were generally supported. The temporary residue limits for ETU residues recommended in 1974 were maintained as guideline levels pending further data. Recommendations for ethylene-diamine limits were withdrawn in view of the proven unsatisfactory nature of the analytical method and its lack of applicability to the required purpose. It was agreed that the well-established $CS_2$ evolution methods remain the only pragmatic methods for the general determination of residues of dithiocarbamates and temporary MRLs were set on that basis. It is realized, of course, that they are non-specific as regards the actual compounds present but nonetheless such methods offer the best currently available procedure for estimating the total dithiocarbamate residue of any sample. It is also recognized that specific methods for determining individual EBDC compounds are unlikely to be evolved.

Very little further useful information concerning the chemistry and residue status of thiram, ziram or ferbam was available to the Panel. Propineb was reviewed for the first time (see Section 4.25.3).

The Meeting noted the availability of a method of analysis, recently adopted as an AOAC official interim first action method, which determines not only free ETU but any potential ETU from unchanged EBDCs or any other ETU precursors.


## 4.25.1    DIMETHYL DITHIOCARBAMATES:  FERBAM, ZIRAM;  THIRAM

### Toxicology

The dimethyl dithiocarbamates possess similar metabolic and toxicologic potential where structures are similar.  Thus, ferbam and ziram were considered to be sufficiently alike to be grouped together.  The concerns relating to reactions with nitrite to form nitrosamines, and the potential for teratogenicity expressed by previous meetings were considered.  The present Meeting determined that an ADI for man could be established.

In the case of thiram, the difference in structure was sufficient to warrant separate consideration.  Thiram is also teratogenic at high dose levels.  In addition, data indicated  that a diet supplying 0.015 mg thiram daily to rats resulted in weight loss, nervous disturbances and severe anaemia.  These data were available in summary form only.  No haematological data were available to the Meeting.  Since in other studies much higher dose levels did not result in significant weight gain changes, the Meeting determined that the no-effect level should be derived from the more extensive data available in a long-term study.  However, concerns relating to haematology were deemed to require investigation and, hence, a temporary acceptable daily intake for man was established for this compound.


### TOXICOLOGICAL EVALUATION

a)    FERBAM *

#### Level causing no toxicological effect

Rat:   250 mg/kg in the diet, equivalent to 12.5 mg/kg bw

Dog:    5 mg/kg bw


#### Estimate of Acceptable Daily Intake for Man

0-0.02 mg/kg bw

b)    ZIRAM *

#### Level causing no toxicological effect

Rat:   250 mg/kg in the diet, equivalent to 12.5 mg/kg bw

Dog:    5 (mg/kg bw)/day


#### Estimate of Acceptable Daily Intake for Man

0-0.02 mg/kg bw

c)   THIRAM *

<u>Level causing no toxicological effect</u>

Rat:   48 ppm in the diet, equivalent to 2.5 mg/kg bw

Dog:   5 (mg/kg bw)/day

<u>Estimate of a Temporary Acceptable Daily Intake for Man</u>

0-0.005 (mg/kg bw)/day

4.25.2    ETHYLENE BISDITHIOCARBAMATES:  NABAM; MANCOZEB, MANEB, ZINEB

<u>Toxicology</u>

a)   NABAM *

Nabam is sodium ethylenebisdithiocarbamate.  Like other fungicides of this class,
it decomposes and is degraded to ETU and ethylenediamine (EDA).  It was evaluated
by previous Joint Meetings and a temporary ADI for man was allocated in 1974.

Nabam is the most acutely toxic of this class of compounds and is goitrogenic.
The Meeting was concerned at the relatively high acute toxicity of this compound and
as it was unaware of any work currently in progress, it was unable to recommend an ADI
for man or to recommend the retention of the existing temporary ADI for man.

However, it is recognized that nabam is used in agriculture only in conjunction
with zinc sulphate to form zineb before spraying.

b)   MANCOZEB *, MANEB * AND ZINEB *

The Meeting was aware of further work in progress and for this reason decided to
retain the existing temporary ADI for man for a period of two years at the present
level of 0-0.005 mg/kg.  This ADI for man applies to mancozeb, maneb and zineb in-
dividually or as the sum of any or of all of these.

4.25.3    PROPYLENEBISDITHIOCARBAMATE:  PROPINEB ZINC PROPYLENEBISDITHIOCARBAMATE
                                                    POLYMERIC

<u>Residue and Analytical Aspects</u>

Propineb is a broad spectrum propylenebisdithiocarbamate fungicide marketed worldwide for pre-harvest treatments of fruits and vegetables.

Residue data provided from various supervised trials of propineb applications
ranged from up to 10 mg/kg on grapes and blackcurrants to less than 0.05 mg/kg in
banana pulp, both expressed as propineb.

The distribution and metabolism of propineb and the associated propylenethiourea (PTU) have been studied.  Propineb and PTU remain mainly on the surface while some other metabolites can penetrate the skins.  PTU is degraded rapidly to yield the same metabolites as does propineb.

Residue analysis is by $CS_2$ evolution and a gas-chromatographic method for ethylenethiourea  (ETU) has been adapted to include PTU.  As for other dithiocarbamates the limits are for residues determined and expressed as $CS_2$.

## Toxicology

Propineb, a propylene analogue of zineb, has moderate acute toxicity.  It degrades to propylene thiourea (PTU).

Orally administered propineb is readily excreted via urine and faeces.  Propineb has a goitrogenic effect.  Following oral administration, high tissue concentration in, and enlargement of, the thyroid and pituitary glands occurred.  These effects are reversible.

A dominant lethal test in mice was negative.  Reproduction studies in rats show no adverse effect at 20 and 60 ppm, but muscular damage (myasthenia) and reduced fertility occurred at 200 ppm.  In a teratogenicity study, teratogenic effect was produced only at a dose which was highly toxic and lethal to the dams.  In a long-term study, a significant increase in the incidence of thyroid benign tumours and skeletal muscle degeneration was observed at dietary level of 1 000 ppm and above.  Significant thyroid enlargement occurred in the males at dietary levels of 100 ppm and above.  In another two-year feeding experiment with rats at a level of 100 ppm, serum protein-bound iodine (PBI) was lowered.

Two studies in dog were carried out (four months and two years).  No adverse effects were seen at 3 000 ppm.  In a study with rats for two years, a dose dependent reduction of growth and significant increase in liver and kidney weights were noted at 100 ppm and above.

In a four-week study with hens, no effects were seen at 1 000 ppm and a dose-dependent increase in thyroid weight occurred at 2 000 – 8 000 ppm.  No neurotoxicity was observed.

Because of the concern of the Meeting regarding the potential for thyrotoxicity and tumorigenicity of PTU (known to occur with ETU), the Meeting recommended a temporary ADI for man.

## TOXICOLOGICAL EVALUATION

### Level causing no toxicological effect

Rat:     10 mg/kg in the diet, equivalent to 0.5 mg/kg bw

Dog:  3 000 mg/kg in the diet, equivalent to 75 mg/kg bw

### Estimate of a Temporary Acceptable Daily Intake for Man

0-0.005 mg/kg bw

## 4.26 DODINE *

### Residue and Analytical Aspects

This compound was re-evaluated in the light of further information requested by the 1974 Joint Meeting and the decision of the 9th Session of the Codex Committee on Pesticide Residues (1977) to increase the maximum residue limits in apples and pears to 5 mg/kg.

Re-examination of the data supports the need for a maximum residue limit of 5 mg/kg in apples and pears based on a seven-day pre-harvest interval. Residues in apples were found mainly in the peel and consisted only of the parent compound.

## 4.27 ETHEPHON (2-chloroethylphosphonic acid)

### Residue and Analytical Aspects

Ethephon is a growth regulator used in liquid formulations on a variety of fruits, nuts and vegetables. It is absorbed and translocated in plants. Radio-tracer studies show that the compound is metabolized in plants to ethylene, the active principle, and phosphate and chloride ions. However, the parent compound itself is sufficiently persistent to require residue limits for fruits, nuts and vegetables in commerce. 14-C-labelled ethephon has been used in tracer studies on 11 crops and soil. It degrades in soil by the same pathway exhibited in plants.

Animal feeding studies with 14-C-labelled ethephon show that it is rapidly eliminated from the body as parent, ethylene, and $CO_2$. Although there is some concentration of residues in feed byproducts, there are no secondary residues in meat, milk, or eggs from feeding those commodities.

Ethephon residues are determined by a gas chromatographic method which has been validated in government laboratories. The lower limit of determination is about 0.05 mg/kg in crops. Residues have been reported in some commodities in commerce.

Extensive data from supervised residue trials were available to the Meeting. In common with other growth regulators, residues tend to be highly variable. Sufficient residue data were available to record guideline limits, in the absence of an ADI for man.

### Toxicology

No toxicological information on ethephon was made available to the 1977 Joint Meeting. Therefore, no toxicological evaluation could be accomplished and no acceptable daily intake for man was established.

4.28 ETHIOFENCARB (2-ethylthiomethylphenyl methylcarbamate)

<u>Residue and Analytical Aspects</u>

Ethiofencarb is a systemic insecticide used especially against sucking insects on vegetables, fruit, field crops and tobacco. It is effective against some organophosphorus-resistant strains. It is formulated as granules for use at sowing or planting time and as emulsifiable concentrates. It is used in various European and in other countries.

In plants and soil ethiofencarb (I) is oxidized. Its sulphoxide (II) and sulphone (III) are also hydrolyzed. In plants and soil the main intact carbamates are II and III. Little or no ethiofencarb is usually detectable at harvest. In animals, the corresponding phenols and their conjugates are the main metabolites. Half-lives of I + II + III in soil vary between two and 12 weeks.

On the basis of feeding diets containing 50 ppm of I to cows, pigs and hens, it can be estimated that residues of I + II + III are unlikely to exceed 0.02 mg/kg in their meat, or in eggs or milk.

Boiling broad beans reduced carbamate residues of 0.13 mg/kg to below 0.02 mg/kg. Cooking the green parts of beans containing 14 to 30 mg/kg of I + II + III reduced the residues to 3 - 7% of the original content. 0.4 - 3.4% of the I + II + III was found in the water used for boiling. Deep and shallow frying of potatoes fortified with 1 mg ethiofencarb/kg decreased the total carbamate residue by about 50%. Baking potatoes caused no appreciable loss of carbamate residues.

Gas-chromatographic procedures for the determination of residues of I + II + III have been elaborated. They are suitable for regulatory purposes, in various crops, eggs, milk and soils.

On the basis of supervised trials, temporary MRLs were recommended. They refer to the sum of ethiofencarb, its sulphoxide and its sulphone expressed as ethiofencarb.

<u>Toxicology</u>

Ethiofencarb is almost completely absorbed in mammals and it is excreted rapidly in the form of metabolites mainly in the urine. The major metabolic reactions are demethylation and subsequent hydrolysis of the carbamate and oxidations to sulphoxides and sulphones. Ethiofencarb inhibits the activity of acetylcholinesterase only transiently.

In a four-week study daily oral administration of 10 mg/kg to male rats led to a maximal depression of 25% of the plasma-cholinesterase after two hours, and of 15% of the erythrocyte-cholinesterase. Recovery occurred almost completely within 24 hours. This dose level was taken as a basis for establishing an acceptable daily intake for man. The fact that at 10 mg/kg marginal effects occurred was taken into account by using a safety factor higher than is usual for cholinesterase-inhibiting compounds.

The compound was not mutagenic in a dominant lethal test, nor was it teratogenic. No three-generation study was available. However, such a study was felt to be important since the depression of the cholinesterase-activity manifests itself often unfavourably in a reduced viability of the pups.

In a two-year feeding study in dogs no untoward effects were seen up to 1 000 ppm. A long-term feeding study in rats revealed no more than a moderate increase of the relative liver weights at 3 000 ppm.

TOXICOLOGICAL EVALUATION

<u>Level causing no toxicological effect</u>

Rat:   10 (mg/kg bw)/day

Dog:  1 000 mg/kg in the diet, equivalent to 25 mg/kg bw

<u>Estimate of a Temporary Acceptable Daily Intake for Man</u>

0-0.1 mg/kg bw

## 4.29 FENAMIPHOS *

<u>Residue and Analytical Aspects</u>

Data received in response to the request from the 1974 Joint Meeting indicated that the residue limit for potatoes should be confirmed but that a higher limit was required for oranges.

## 4.30 FENBUTATIN OXIDE (bis/tris(2-methyl-2-phenylpropyl)tin/ oxide)

<u>Residue and Analytical Aspects</u>

Fenbutatin oxide is used as an acaricide as a foliar application on a considerable scale and in various countries on a wide variety of fruits and fruiting vegetables.  It is mainly marketed as a 50% wettable powder, but also as a suspension concentrate. The rates of application vary depending on crop, pest and method of application;  normal rates are usually 1 - 2.5 kg a.i./ha.  Recommended pre-harvest intervals in various countries vary considerably for one and the same crop.

Residue data from supervised trials were obtained from several countries.  Residues do not penetrate through the skin of treated fruit to any great extent, and are, therefore, generally below the limit of determination in the pulp.

Extensive data are available on the fate of the residues in plants and on the levels of residues in products of animal origin arising from the feeding of treated commodities including the pulp of treated fruits.  On apples, after 33 days the remaining residue consists of about 85% unchanged parent compound, 5% of the principal organotin metabolite 1,1,3,3-tetrakis($\beta,\beta$-dimethylphenethyl)-1,3-dihydroxydistannoxane (SD 31723) and about 10% of inorganic tin.

The degradation of the main metabolite SD 31723 is faster than that of the parent compound and thus, under practical conditions, the metabolite will generally not form an appreciable part of any residue present.

Occasionally residues not exceeding 0.2 mg/kg of the parent compound could be found in the kidneys of cows fed with treated feeds; in meat and milk no residues were detectable (limit of determination 0.02 mg/kg).

Under field conditions fenbutatin oxide applied to the soil remains in the upper layers and is slowly degraded, forming non-mobile metabolites. Washing fruits and vegetables removes 60% or more of the original residue, and removal of the outer peel of apples or citrus reduces the residue by at least 75%.

Gas-chromatographic methods for residue analysis for fenbutatin oxide on fruit crops and fruit of vegetables are available, which are suitable or can be adapted for regulatory purposes. The methods have been adapted for residues in milk and animal tissues. Methods for the determination of the principal metabolite have also been developed.

MRLs were recommended for various crops, animal feed and liver and kidney of live-stock animals. They refer to fenbutatin oxide, excluding any metabolites.

## Toxicology

Fenbutatin oxide is moderately acutely toxic by the oral route, producing severe gastrointestinal irritation. Acute oral administration to rabbits reduced spermatogenesis and resulted in multi-nucleated spermatid production. It is excreted rapidly, mostly unchanged in the faeces, indicating that it is poorly absorbed. In the rat it is metabolized to $\beta,\beta$-dimethylphenethylstannoic acid (SD 33608) and 1,1,3,3-tetrakis ($\beta,\beta$-dimethylphenethyl)-1,3-dihydroxydistannoxane (SD 31723). The latter metabolite and fenbutatin oxide failed to produce oedema of the central nervous system of the rat at 100 and 1 000 mg/kg respectively. Fenbutatin oxide produced no genotoxic effects in several _in vitro_ and _in vivo_ systems. Adequate short- and long-term studies are available in rats and dogs as well as a teratology study in rabbits, a multi-generation reproduction study in rats, and a carcinogenicity study in mice. No effect levels were demonstrated in rat and dog studies and an acceptable daily intake for man was established on the basis of these studies. The no-effect level was established as 15 mg/kg because of emesis and diarrhoea in dog, and because of the known toxic effects of organotin compounds on the brain a safety factor of 100 was used.

## TOXICOLOGICAL EVALUATION

### Level causing no toxicological effect

Rat: 50 mg/kg in the diet, equivalent to 2.5 mg/kg bw

Dog: 15 (mg/kg bw)/day

### Estimate of Acceptable Daily Intake for Man

0-0.03 mg/kg bw

## 4.31 FENITROTHION *

### Residue and Analytical Aspects

The fate of fenitrothion applied to various grains was evaluated in 1974 and 1976. Further studies on a variety of grains including barley, oats, rice, sorghum and wheat have revealed that the rate of decline in residue levels in stored raw cereals can be predicted from a knowledge of the temperature, relative humidity and residue concentration.

These data, taken in conjunction with information contained in previous monographs, have provided adequate assurance that the maximum residue limits already recommended for rice, in the husk and hulled, and wheat can be extended to barley, oats, sorghum, rye and other raw cereals.

Additional studies following the application of fenitrothion to malting barley have shown that there is no adverse effect from the treatment on the quality of the malt produced.

The previously recommended limit for fenitrothion on rice and wheat was amended to include sorghum and other raw cereals.

### Toxicology

Additional information on absorption, distribution, excretion, potentiation, teratogenicity and neurotoxicity has become available.

Fenitrothion is easily absorbed from the gastro-intestinal tract and distributed into various tissues.  The main metabolites are 3-methyl-4-nitrophenol and O-desmethyl-fenitrothion oxygen analogue which are eliminated rapidly, predominantly by the kidneys. Tissue residues are very low.  There are some indications that fenitrothion induces increased ocular pressure after oral application in both dogs and rats.  However, this can be expected since it depresses cholinesterases.  No-effect levels in both rats and dogs were demonstrated at 5 ppm in the diet.  There were no reasons to alter the previously established ADI for man.

### TOXICOLOGICAL EVALUATION

### Level causing no toxicological effect

Rat:  5 mg/kg in the diet, equivalent to 0.25 mg/kg bw

Dog:  5 mg/kg in the diet, equivalent to 0.125 mg/kg bw

### Estimate of Acceptable Daily Intake for Man

0-0.005 mg/kg bw

## 4.32 FENTHION *

### Residue and Analytical Aspects

Following evaluation by the Joint Meeting in 1971 a request was made for additional information on the occurrence of residues in citrus fruit, coffee, cucurbits, onions, potatoes, sugar beet roots and tops.  The 1977 Session of the Codex Committee on Pesticide Residues requested a review of the MRLs in the light of information provided by governments and other relevant data such as disappearance.

Results from a considerable number of supervised residue studies were provided by the principal manufacturer and from Poland, the Philippines and Turkey.  Other inform-. ation was provided from USA and published scientific literature.

One extensive study from Poland carried out over two seasons demonstrates the rate of decline of fenthion residues in apples, cherries and plums.  There is considerable similarity between the rates of decline of residues on the different species of fruit that is partly influenced by growth dilution.  Fourteen or 21 days after spraying, the difference in residue level between two distinctly different fruits is often less than the difference between two cultivars of the same species or between residue levels in the same cultivars in two succeeding years.  This study is a very useful indicator of the variability of residue levels notwithstanding efforts to standardize conditions and test procedures.  It points to the danger in placing too much reliance on limited amounts of data as representing the position in different regions, seasons, cultivars, formulations and application techniques.

Residue data from studies in Turkey and Japan point to the need to raise the maximum residue limit for fenthion in citrus fruit where residues occur mainly in the peel. The Japanese studies show that the use of fenthion on potato plants does not give rise to measurable residues in potato tubers.

Further studies have confirmed that fenthion present in livestock feed or used for the control of external parasites gives rise to low but transitory residues in milk. The metabolism of fenthion in cattle has been clarified by these studies.  Further studies have also shown that sunlight causes rapid degradation of fenthion and the production of polar products believed to be polymeric phenols.

The above mentioned new data confirmed that the MRLs previously recommended are adequate and are not excessively high for residues resulting from approved uses on apples, cherries, grapes, peaches and rice.  The information considered also provided a basis for recommending additional maximum residue limits for bananas, beans, onions, pears, plums, potatoes, strawberries and tomatoes.

Extensive studies in Turkey, however, indicate that there is a high risk that olives treated for the control of fruit flies will contain residues above the maximum residue limit even when a 30-day pre-harvest interval is observed.  Furthermore, no information was received to assist in resolving the questions about residues in coffee, cucurbits and sugar beets.

With the exception of the limit for citrus the MRLs previously recommende were, therefore, confirmed and new recommendations were made for some additional commodities.

These limits refer to the sum of fenthion, fenthion sulphoxide, fenthion sulphone and the oxygen analogues of these compounds, determined as the sulphone of the oxygen analogue and expressed as fenthion.


## Toxicology

No new toxicological data were made available in sufficient time, and toxicological consideration of this pesticide was deferred to a future meeting.


## 4.33 HEPTACHLOR *


## Residue and Analytical Aspects

At the 9th Session of the Codex Committee on Pesticide Residues (1977), a question was raised regarding the relationship between the Extraneous Residue Limit (ERL) of 0.05 mg/kg for sugar beet and the earlier established ERLs (then called Practical Residue Limits) of 0.15 mg/kg in milk and milk products (fat basis) and 0.2 mg/kg in carcass meat (fat basis) arising from the use of sugar beet tops and leaves and wet and dry pulp in rations for cattle.

The Joint Meeting was aware of a considerable amount of data that may already be available in the open literature that would assist in answering the question. Because it was impossible to review this in the time available, the matter was deferred to a future meeting.


## 4.34 IMAZALIL (1-($\beta$-allyloxy-2,4-dichlorophenethyl)imidazole)


## Residue and Analytical Aspects

Imazalil is a systemic fungicide which is effective against a large numger of fungi. It is formulated as a seed dressing for cereals, an emulsifiable concentrate and a wettable powder. The main pre-harvest applications are on cucurbits and bananas. Further uses are being developed.

Imazalil is also effective against fungal infections on citrus fruit and bananas after harvest.

In plants $\alpha$-(2,4-dichlorophenyl)-1H-imidazole-1-ethanol and other unidentified polar, hydrophilic compounds are the main degradation products. The half lives of imazalil on oranges and in soil were 12 - 20 weeks and four - five months respectively.

No significant differences in the residue levels on citrus fruits which were stored for 32 days at $4^\circ$ C and 22 - $25^\circ$ C were noticed.

Gas-chromatographic methods for the determination of imazalil are suitable for regulatory purposes.

A number of temporary MRLs were recommended. They refer only to residues of imazalil.

<u>Toxicology</u>

Imazalil is moderately acutely toxic to mammals.  Preliminary studies indicate that it is rapidly absorbed, metabolized and excreted by the rat.  The major metabolites are a-(2,4-dichlorophenyl)-1H-imidazole-1-ethanol and 2,4-dichloromandelic acid.  Dominant lethal mutagenic effects were not evident in male and female mice.  No teratological effects were noted in rats and no reproductive effects were observed in a three generation rat study.  However, survival of pups at weaning was adversely affected at levels higher than 5 mg/kg.  Long-term studies in the rat and the study in the dog were inadequate because of unresolved questions concerning the rat kidney and dog liver pathology.  Because of these concerns, only a temporary ADI for man was established for this compound based on short-term studies.

TOXICOLOGICAL EVALUATION

<u>Level causing no toxicological effect</u>

Rat:     5 (mg/kg bw)/day

Dog:   1.25 (mg/kg bw)/day

<u>Estimate of Temporary Acceptable Daily Intake for Man</u>

0–0.01 mg/kg bw

4.35 IPRODIONE (3-(3,5-dichlorophenyl)-<u>N</u>-isopropyl-2,4-dioxoimidazolidine-1-carboxamide)

<u>Residue and Analytical Aspects</u>

Iprodione is used against a relatively broad range of fungus diseases on a wide range of fruits and vegetables.

Its use is authorized, or is in course of registration, for various crops and countries.  It is marketed in the form of a wettable powder, a suspension concentrate and an emulsifiable concentrate.  The products are mainly used as a spray on the aerial parts of growing crops, for post-harvest dipping of fruit, as a dip for seed potatoes and as a seed treatment.  Application rates vary according to the crop/disease situation and regional conditions.

Residue data were obtained from supervised trials carried out in various countries with different climatic conditions.  Studies with unlabelled and $^{14}$C-phenyl-labelled products showed that iprodione does not appreciably penetrate through the plant cuticle. The residue on the surface of the plants had a half-life of about 30 – 60 days.  It was converted into 1-(3,5-dichlorophenylcarbamoyl)-3-isopropyl-2,4-dioxoimidazolidine (1-(3,5-dichlorophenylcarbamoyl)-3-isopropylhydantoin, RP 30228), which represented up to 35% of the remaining residue.  Two – 5% of this residue consisted of minor degradation products, including 1-carbamoyl-3-(3,5-dichlorophenyl)hydantoin (RP 32490).

Wheat and strawberry plants grown on soil treated with iprodione took up small amounts of the compound (equivalent to 0.7 – 1.3% of the total applied to the soil surface).  Within the plant the parent compound was converted into the metabolite RP 30228, small amounts of RP 32490 and some more polar unidentified products.

The degradation of residues in soil follows the same pattern.  The half-life at initial levels of 2 and 5 mg/kg is about 30 days.  After 12 months incubation under aerobic conditions, no more than 3% of the remaining residue is in the form of unchanged iprodione.  The parent compound is only slightly mobile, remaining mainly in the upper 0 - 15 cm layer.  The metabolite RP 30228 is less soluble in water than the parent compound (0.5 mg/l compared to 13 mg/l) and virtually all remained in the 0 - 5 cm layer.  Residues in wine were approximately 15 - 25% of those on the harvested grapes.

Gas-chromatographic methods using electron capture detectors have been developed for the analysis of residues in several fruits and vegetables must and wine which are suitable or can be adapted for regulatory purposes.  The limit of determination is about 0.01 - 0.02 mg/kg.  No loss of residue was found over more than one year at -18° C.

MRLs for iprodione on various fruits and vegetables were recommended.  They refer to iprodione, excluding any metabolites.

## Toxicology

Iprodione is readily absorbed and rapidly excreted mainly as metabolites with an intact hydantoin moiety.  The compound was not teratogenic.  In a 3-generation study in rats, there was a slight but statistically significant reduction in postnatal growth at 2 000 ppm.  This effect was only marginal at the lower dose of 500 ppm which is regarded as a no adverse effect level.  In a short-term study in dogs no major effects occurred up to 2 400 ppm.  Likewise long-term studies in mice and rats revealed no effects up to 1 250 ppm.  No ocular alterations were found in any study.

## TOXICOLOGICAL EVALUATION

### Level causing no toxicological effect

Mice:  1 250 mg/kg in the diet, equivalent to 160 mg/kg bw

Rat:     500 mg/kg in the diet, equivalent to  25 mg/kg bw

### Estimate of Acceptable Daily Intake for Man

0-0.3 mg/kg bw

## 4.36 LINDANE *

### Residue and Analytical Aspects

The ADI for man and recommended MRLs were temporary and due for review in 1977.  Further information on residues in animal feedstuffs in relation to residues in milk, etc., was listed as desirable by the 1973 and 1974 Meetings.  Further residue data for various crops showed that the residue levels were at or below the limits recommended by the 1975 Meeting.

As no other information on these crops had become available, the limits recommended by the 1975 Meeting  and those of the 1972 Meeting  which were not changed in 1975, were maintained but, consequent to the establishment of an ADI, would no longer be temporary.  Additional MRLs and an extraneous residue limit were also recommended.

<u>Toxicology</u>

In view of the currently very low intake of lindane from a wide variety of sources, the relative rapidity with which it is degraded in mammalian organisms and in the environment to much less acutely toxic compounds, and the relatively high levels required to produce adverse effects in man and laboratory animals (      especially carcinogenesis), there appear to be no further toxicological objections to the continued maintenance of the previously established ADI for man.

Neither of the new mouse studies reported here nor the new rat study both undertaken with lindane of much greater purity than previously tested provide any evidence of tumour or cancer induction following exposure to lindane.  Since the information required by the 1974 Meeting has now been satisfied, the present Meeting decided that the previously allocated ADI for man is no longer temporary.

TOXICOLOGICAL EVALUATION

<u>Level causing no toxicological effect</u>

Rat:   25   mg/kg in the diet, equivalent to 1.25 mg/kg bw

Dog:   1.6 (mg/kg bw)/day

<u>Estimate of Acceptable Daily Intake for Man</u>

0—0.01 mg/kg bw

4.37 MALATHION *

<u>Residue and Analytical Aspects</u>

Malathion was last evaluated by the Joint Meeting in 1975.  Further information was desirable on residues in stored grains.

It can be concluded from the 1976 evaluation of fenitrothion and a previous report on supervised trials that a level of 2 mg/kg in whole meal and white flour could be exceeded.  The Meeting was informed of other data which also suggested that a limit higher than 2 mg/kg was likely to be required.  It was decided to defer a decision pending a full review of the subject at a future meeting.  A MRL for bran was recommended.

4.38 MALEIC HYDRAZIDE *

<u>Residue and Analytical Aspects</u>

Data received in response to a request from the 1976 Meeting showed that residues of maleic hydrazide in potatoes were stable to cooking;  the guideline levels were maintained.  Data on residues in tobacco enabled guideline levels to be recorded.

## Toxicology

No new toxicological information has been submitted to the Meeting.  It was previously determined that the carcinogenicity potential could not be assessed adequately and since the long-term rat feeding study was on a small number of animals, no toxicological evaluation has been attempted for the sodium salt of maleic hydrazide. Data are totally inadequate for toxicological evaluation of the diethanolamine salt. Since the concerns of the 1976 Joint Meeting have not yet been addressed it was not possible to complete a toxicological evaluation or to establish an ADI for man.

## 4.39 METHIDATHION *

## Residue and Analytical Aspects

At the 9th Session of the Codex Committee on Pesticide Residues (1977), it was stated that the recommended maximum residue limit for leafy vegetables was insufficient to cover also the use of this compound on glasshouse crops.  No information was available to indicate whether or not the residues resulting from the use of methidathion on glasshouse crops are covered by the recommended MRLs of 0.2 mg/kg.

## Toxicology

No further toxicological information was received and no further action was taken.

## 4.40 METHOMYL *

## Residue and Analytical Aspects

The Netherlands submitted additional data on use patterns in tomatoes and bell peppers and residue data on tomatoes;  residues were generally low, 0.01 - 0.06 mg/kg, at the pre-harvest interval of 0 - three days.

None of the requirements of the 1975 and 1976 Joint Meetings have been met.  The guideline levels previously recorded are maintained.

## Toxicology

Although toxicological studies on methomyl were received during the Meeting they could not be considered, as full data were not available.  Thus, the Meeting felt unable to evaluate these studies.

Therefore, no ADI for man was established.

## 4.41 METHOXYCHLOR *

## Toxicology

Several new studies were examined by the Meeting.

Methoxychlor is mainly degraded by hydrolysis of the methyl ether leading to a polar phenol which is rapidly excreted.

Methoxychlor does not exhibit a teratogenic potential in rats.

Several carcinogenicity studies in rats and mice at dose levels up to 3 500 ppm were negative.  Therefore, the ADI for man established in 1965 remains unchanged.


TOXICOLOGICAL EVALUATION

<u>Level causing no toxicological effect</u>

Rat:  200 mg/kg in the diet, equivalent to 10 mg/kg bw

<u>Estimate of Acceptable Daily Intake for Man</u>

0-0.1 mg/kg bw


4.42 PHORATE (<u>OO</u>-diethyl <u>S</u>-ethylthiomethyl phosphorodithioate)

<u>Residue and Analytical Aspects</u>

Phorate is a widely used soil insecticide.  Maize, cotton, cereal grains and vegetables are treated usually at the time of planting, but sometimes by broadcast treatments of granules or foliage application to the growing crop.  Residues in harvested commodities do not exceed 0.1 mg/kg.  Carrots are capable of taking up and retaining quantities of phorate but most of the residue is in the peel or in the crown of the carrot which is generally removed prior to consumption.

Soil treatments lead to higher residues than do foliage treatments, largely because the compound is relatively stable in the soil and is readily taken up by the plant's root system.  In animals phorate is oxidized at both thiono and thioether sites and hydrolized.  No significant amounts of phorate or its metabolites were found in milk, eggs or tissues of livestock receiving phorate at levels likely to occur in animal feeds.

The metabolism of phorate in plants is largely similar to that occurring in animals. Phorate metabolites are relatively stable in storage and processing.  At rates recommended in good agricultural practice, there appears to be very little likelihood of the carry-over of residues in soil affecting subsequent crops.

A method of analysis based on the oxidation of all the phorate-related residues to the oxygen analogue of phorate sulphone followed by measurement by gas-liquid chromatography is suitable or can be adapted for regulatory purposes.

In the absence of an ADI, guideline levels were recorded.  They refer to the sum of phorate, phorate sulphoxide, phorate sulphone, and the oxygen analogues of these compounds, determined as the sulphone of the oxygen analogue and expressed as phorate.

<u>Toxicology</u>

Less than 40% of a 2 mg/kg oral dose of phorate to rats was excreted in six days, and the brain, liver and kidney contained unextractable residues. The major fully esterified metabolites in both plants and mammals are sulphoxides and sulphones of phorate and of its oxygen analogue, phoratoxon;  these have greater anticholinesterase activity than phorate.  Hydrolysis of phorate itself is a minor metabolic pathway in mammals.

The biochemical data available on phorate are inadequate to determine if the oxidative metabolites are retained in mammalian tissues or organs.  Toxicological data submitted were insufficient, except for cholinesterase inhibition, to determine the short- and long-term effects of phorate and its oxidative metabolites.  Although the Meeting considered setting a temporary ADI for man this could not be established in the absence  of long-term studies.  Further studies are required to evaluate:

(a)    carcinogenic potential;

(b)    teratogenic potential;

(c)    potential neurotoxicity;  and

(d)    toxicity of metabolites.

Observations in human subjects are also desirable.


4.43 PHOSMET (<u>OO</u>-dimethyl <u>S</u>-phthalimidomethyl phosphorodithioate)

<u>Residue, Analytical and Toxicological Aspects</u>

No information on phosmet was available to the Meeting.  Therefore, no toxicological or residue evaluation could be undertaken.


4.44 PIRIMIPHOS-METHYL *

<u>Residue and Analytical Aspects</u>

In 1976 a request was made for results from commercial trials with pirimiphos-methyl in further commodities and the Codex Committee on Pesticide Residues at its 9th Session (1977) sought clarification of the intake by humans, the apparently high level of residues in bran and the distinction between commodities destined for animal feed and human food.

Additional information concerning the level and fate of pirimiphos-methyl residues on sorghum, maize and in milled products from wheat has become available.

The rate of application of pirimiphos-methyl post-harvest for grain protection depends upon the degree of protection desired, the length of storage, the temperature and humidity of the grain and the anticipated period of storage.  At 25$^\circ$ C and low relative humidity the half-life of pirimiphos-methyl on grain is of the order of 80 weeks.  Under hot humid conditions, the half-life is comparatively short and, therefore, higher initial deposits are required to achieve acceptable protection against insect damage.

A model has been developed whereby it is possible, given the temperature, relative humidity and proposed application rate, to predict the storage life of the deposit and to estimate the residue level at any desired date. Residue data from extensive pilot and commercial trials have been compared with the above model and there has been remarkably good agreement.

The fate of residues of pirimiphos-methyl on grain subjected to milling has been revealed in several studies. These show that the bulk of the residue is removed in the bran and that only about 10% of the amount present on the grain finds its way into bread. It was recognized that small amounts of bran are used for dietetic purposes but it was considered that the amount contributed to the intake from this source was acceptable because only occasional batches would contain residues approaching the recommended MRLs. (See also Section 3.5).

Attention was drawn to the importance of the extracting solvent and extraction technique on the recovery of residues from raw grains, milling products and cooked cereals.

The existing MRLs for several individual cereals were replaced by a single limit for raw cereals.

<u>Toxicology</u>

Since the 1976 Joint Meeting, no new toxicological information has become available and no further action was taken.

4.45 PROPARGITE (2-(4-<u>tert</u>-butylphenoxy)cyclohexyl prop-2-ynyl sulphite)

<u>Residue and Analytical Aspects</u>

Propargite is an acaricide which has been used on a variety of food crops since 1967. The technical grade product has a minimum active ingredient content of 85%. Information on the impurities was available to the Meeting. It is commercially available in formulations of emulsion concentrates, wettable powders and dusts. The product has worldwide use, and national tolerances have been established in at least four countries.

The residues on harvested commodities are primarily on the surface and consist of propargite <u>per se</u>. Available data indicate that the route of degradation is through hydrolysis to 2-(4-<u>tert</u>-butylphenoxy)cyclohexanol and propargyl alcohol. The two hydrolysis products have not been detected on harvested crops by chemical methods at a minimum detection level of 0.1 mg/kg. Their absence is attributed to volatility. The conclusions on the fate of residues in plants are based on an analytical method for postulated degradation products; no radiotracer studies were available.

The fate in animals follows a metabolic pathway initially similar to that in plants, namely hydrolysis to the glycol ether and propargyl alcohol. Further degradation to butyl phenol and cyclohexanediol probably occurs, followed by excretion as conjugates. Some unchanged propargite occurs in milk, body fat and in the liver as a transient residue. The significant difference in metabolism from that of the related compound "Aramite" may be in the initial hydrolysis product propargyl alcohol as compared with $\beta$-chloroethanol released by Aramite.

Propargite is used on a number of crops which yield byproducts used in commercial animal feeds. Data on controlled feeding studies with dairy cattle, poultry and hogs were available. After considering the probable intake from all feed sources, it was concluded that temporary MRLs could be recommended for milk, eggs, meat fat and poultry fat.

Data from supervised trials on many crops were available to the Meeting. It was concluded that the available data were sufficient to support the temporary MRLs recorded in Annex I.

## Toxicology

Propargite is readily absorbed from the gastrointestinal tract and is excreted rapidly as metabolites which are formed mainly by hydrolysis of the sulphite ester.

After feeding a diet containing 20 ppm of propargite the residues in milk or in tissues were of the order of 0.1 mg/kg.

The compound is not teratogenic in the rat.

In a 3-generation reproductive study no adverse effects were revealed at concentrations up to 300 ppm. This concentration is, therefore, regarded as a no-effect level.

In a 2-year feeding study with up to 900 ppm in dogs which included extensive laboratory and histological investigations, no adverse effects were noted. A 2-year feeding study in rats up to 2 000 ppm showed appreciable but not dose-related variations in the relative liver and kidney weights in all treated groups. However, no major adverse effects were seen. Owing to a poor survival rate and the termination of the study with the high dose animals after 78 weeks, the finding that the tumour incidence was not elevated is of limited value. This study was not considered to fulfil the criteria for an acceptable long-term study. Therefore, a safety factor of 200 was used.

Propargite is structurally related to Aramite, a compound which is suspected of being a carcinogen. Although there are no indications that propargite acts similarly, the need for satisfactory carcinogenicity studies was expressed.


TOXICOLOGICAL EVALUATION

## Level causing no toxicological effect

Rat: 300 mg/kg in the diet, equivalent to 15 mg/kg bw

Dog: 900 mg/kg in the diet, equivalent to 22 mg/kg bw

## Estimate of Temporary Acceptable Daily Intake for Man

0-0.08 mg/kg bw

## 4.46 PROPOXUR *

### Residue and Analytical Aspects

Clarification of the components to be included in the residues of propoxur was sought at the 9th Session of the Codex Committee on Pesticide Residues (1977). The Meeting confirmed that the MRLs refer to:

"the sum of propoxur and its main metabolites, i.e., 2-hydroxyphenylmethylcarbamate and 2-isopropoxyphenyl hydroxymethylcarbamate, expressed as propoxur".

## 4.47 QUINTOZENE *

### Residue and Analytical Aspects

The Meeting discussed questions raised at the 9th Session of the Codex Committee on Pesticide Residues (1977) relating to the composition of quintozene residues and decided that the compounds included in the MRLs for quintozene should be quintozene, pentachloroaniline and methyl pentachlorophenyl sulphide. Hexachlorobenzene and pentachlorobenze should be dealt with separately as extraneous residue limits.

### Toxicology

In two mutagenicity studies, one with bacteria and one with mice no indication of a mutagenic action of quintozene was obtained. At the 1975 Meeting it was found desirable that further work on the influence of quintozene on the formation of subcutaneous fibrosarcomas in female mice should be made avilable. In two carcinogenicity studies with very high dose levels of quintozene no increased numbers of tumours was found in the treated groups. There is no indication, therefore, that administration of quintozene resulted in carcinogenic activity.

### TOXICOLOGICAL EVALUATION

### Level causing no toxicological effect

Rat: 25 mg/kg in the diet, equivalent to 1.25 mg/kg bw

Dog: 30 mg/kg in the diet, equivalent to 0.75 mg/kg bw

### Estimate of Acceptable Daily Intake for Man

0-0.007 mg/kg bw

## 4.48 THIABENDAZOLE *

### Residue and Analytical Aspects

Thiabendazole is now applied to many classes of crops and commodities in pre-plant and pre-harvest as well as post-harvest treatments. Residues resulting from the pre-plant and pre-harvest application of thiabendazole were generally low to very low.

The application of thiabendazole as a seed dressing for cereals and cotton does not give rise to residues in the harvested crop.  The pre-harvest use of thiabendazole on bananas, rice, soybeans and wheat and as a fumigant in glasshouses and similar situations produces only small residues on tomatoes and strawberries.  In the treatment of stored potatoes the method of application has an important bearing on the level of the resulting residue which is largely confined to adhering soil and the outer part of the skin.  Much of this residue is removed by washing.  The amount removable by washing decreases as the potatoes remain longer in storage, but the residue in washed potatoes does not exceed 5 mg/kg.

Up to 99% of the residue present in the unwashed potato is removed by processing and cooking prior to consumption.  Because of the extreme variability in residue levels occurring on unwashed potatoes, as a result of varying soil load, it was recommended that limits should apply only to potatoes which have been washed before analysis.  Following application to sugarbeet entering pile storage, a large proportion of the amount applied becomes bound to the soil coating the beet roots and there is a further loss during storage.  Whilst a significant residue remains in the sugarbeet pulp and in the molasses, no residues remain in the beet sugar.  These studies provide an adequate basis for recommending MRLs in sugarbeet, sugarbeet pulp and sugarbeet molasses.

When thiabendazole is administered to livestock at levels vastly in excess of those likely to occur in animal feeds, significant residues remain in animal tissues but these disappear completely within a few days of withdrawing the thiabendazole.  These new studies serve to confirm the previous recommendations made for an MRL for milk, meat and edible offal of cattle, goats, horses, pigs and sheep at or about the limit of determination (0.1 mg/kg).

Whereas in 1974 it was stated that residues resulting from the use of thiabendazole consisted of thiabendazole and 5-hydroxythiabendazole, it is recognized that this applies only to residues in foods of animal origin such as meat, edible offal, milk and eggs.  An analytical method for measuring both the parent compound and the metabolite in such commodities is available.  Residues in foods of plant origin consist entirely of thiabendazole.  The recommendations were amended accordingly.

In the light of the new and previously available data, additional MRLs were recommended.  The limits for potatoes (unwashed) and potatoes (washed) have been replaced by a recommendation which directs that potatoes should be washed before being analyzed.  The limits refer to thiabendazole.  Limits for commodities of animal origin refer to the sum of thiabendazole and 5-hydroxythiabendazole.

<u>Toxicology</u>

A study in human volunteers with a dose of 250 mg/day corresponding to 3 - 4 mg/kg bw was reviewed.  No effects could be ascribed to the thiabendazole administration, although many parameters were measured including a thyroid function test.

In 1970 an ADI for man was established of 0.05 mg/kg bw based on a no-effect level in the rat of 10 mg/kg bw.  Observations on man allowed the Meeting to increase the ADI for man.

TOXICOLOGICAL EVALUATION

Level causing no toxicological effect

    Rat:  10 (mg/kg bw)/day

    Dog:  20 (mg/kg bw)/day

    Man:   3 (mg/kg bw)/day

Estimate of Acceptable Daily Intake for Man

    0-0.3 mg/kg bw

## 4.49 THIOPHANATE-METHYL *

Toxicology

New data on biochemical aspects were made available. The Meeting reconfirmed the previously established ADI for man of 0-0.08 mg/kg bw.

## 4.50 TRIFORINE (1,4-bis(2,2,2-trichloro-1-formamidoethyl)piperazine)

Residue and Analytical Aspects

No data were available for consideration.

Toxicology

In rats, the compound was rapidly absorbed and excreted mainly in the urine. The main metabolite was 1-(2,2,2-trichloro-1-formamidoethyl)piperazine.

Full toxicological data are required before a recommendation can be made of an ADI for man.

## 5. COMPARISON OF THEORETICAL POTENTIAL DAILY INTAKES OF PESTICIDE RESIDUES WITH THEIR ACCEPTABLE DAILY INTAKES

Maximum residue limits are recommendations for one parameter on which competent authorities can judge the acceptability of foods moving in trade. As they reflect the maximum residues that could result from good agricultural practice they uphold the principle that pest control agents should be applied in such a manner as to leave residues that are the smallest amounts practicable and that are toxicologically acceptable.

On various previous occasions Meetings have pointed out that a critical appraisal of any risk for the health of consumers in specific situations can only be reached by measuring the actual intake in the particular situation and by comparing the results with the ADI for man that has been established. Estimates of such intakes may be obtained from various kinds of monitoring studies including the analysis of individual foods "market baskets", restaurant meals and total diets. Such studies are a matter for national surveillance.

The Meeting recognizes that while the limitations of available analytical methodology and scientific resources make it difficult to obtain comprehensive information on total actual dietary intake of all pesticides from market basket or total diet studies, it is possible to measure the actual intake of individual compounds. The Meeting, therefore, encourages national governments to make studies and expresses the hope that every possible assistance will be given to ensure they can be organized and conducted.

In the absence of results from monitoring studies in many countries, since 1969 calculations of theoretical potential maximum daily intakes in typical diets in selected countries have been undertaken by WHO and the results have been considered at Joint Meetings. These theoretical intakes have been calculated from food consumption figures and MRLs recommended by Meetings. They have included a number of assumptions. For example, the calculations have assumed that the entire supply of any food item is treated with the particular pesticide, that the residue is always present at the level of the MRL, that this maximum level of residue does not decrease during washing, cooking or other processing before consumption and that the consumer eats the food (including any non-edible portions) every day for a life time. Notwithstanding these assumptions, it has been found that few pesticides show a theoretical potential to exceed the ADI for man. Following re-evaluation of some of the results using known residue disappearance data, it was noted that on very few occasions did the theoretical potential daily intake exceed the ADI for man.

The Meeting observed that while this type of calculation has some value in determining priorities for further studies, it was not intended for and was expressly not suitable for determining whether or not to allow the introduction or the extended use of the compound in a given country.

## 6.  FUTURE WORK

The following items should be considered in 1978 or at a future Joint Meeting:

1.  Pesticides scheduled at a previous or at the present Meeting for evaluation or re-evaluation at the 1978 Meeting.

2.  The pesticides aminocarb, benzoximate, tetrachlorvinphos and thiofanox which are still outstanding from Priority List IV of the Codex Committee on Pesticide Residues

3.  Pesticides for which relevant information has become available concerning "Requirements", without accompanying re-evaluation dates being specifically recorded at previous meetings.

4.  The definition and use of the terms "Temporary Acceptable Daily Intake", "Conditional Acceptable Daily Intake" and "Maximum Residue Limit" and the incorporation of possible changes in the Glossary (FAO/WHO, 1976a).

5.  The possible occurrence of nitrosamines as residues following the use of agricultural pesticides, arising from the processes of manufacture, uses of corrosion inhibitors in formulations or from other sources.

# 7.  RECOMMENDATIONS

1.  Parties submitting data to FAO and WHO should be invited to present comprehensive summaries following the established format of monographs;  the summaries should accompany the submission of the detailed data.  In addition, they should present literature searches for additional relevant information concerning the pesticides under consideration.

2.  Because the evaluations made by the Joint Meeting are of great value to regulatory agencies of member countries, FAO and WHO should examine ways and means for assembling data from all available sources and for generally strengthening the activities on the evaluation of pesticide residues.

3.  Significant advances have been made during the past few years, in the science and methodology of toxicology and in the safety evaluation of chemicals.  Therefore, it is highly desirable to convene a special meeting to consider in detail the methodologies of testing and the principles of interpretation of experimental findings for the purpose of assessing the safety of pesticides.

4.  It is recommended that WHO seek the cooperation of the International Registry for Potentially Toxic Substances of the United Nations Environment Programme for the acquisition of toxicological data on pesticides.  Furthermore, in view of the value of observations of the effects of pesticides in man, it is also recommended to seek the cooperation of the World Federation of Associations of Clinical Toxicology Centres and Poison Control Centres and National and International Societies of toxicology as well as other organizations in developing the relevant data banks.

5.  Because of the concern on the total environmental load and human body burden of organochlorine chemicals, WHO should convene a special meeting to consider this phenomenon and the associated potential hazards to human health.

6.  The previously recommended consolidated Table containing all current evaluations and recommendations of Joint Meetings (FAO/WHO, 1977a, pp. 18-19) which has been presented to this Joint Meeting in its draft form should be finalized and published.

## **REFERENCES**

### PREVIOUS FAO AND WHO DOCUMENTATION

FAO/WHO (1958)  —  Procedures for the testing of intentional food additives to establish their safety for use.  Second report of the Joint FAO/WHO Expert Committee on Food Additives.  FAO Nutrition Meetings Report Series, No. 17:  WHO Technical Report Series, No. 144

FAO/WHO (1961)  —  Evaluation of the carcinogenic hazards of food additives.  Fifth Report of the Joint FAO/WHO Expert Committee on Food Additives.  FAO Nutrition Meetings Report Series, No. 29;  WHO Technical Report Series, No. 220

FAO/WHO (1962)  —  Principles governing consumer safety in relation to pesticide residues.  Report of a meeting of a WHO Expert Committee on Pesticide Residues held jointly with the FAO Panel of Experts on the Use of Pesticides in Agriculture.  FAO Plant Production and Protection Division Report, No. PL:1961/11;  WHO Technical Report Series, No. 240

FAO/WHO (1964)  —  Evaluation of the toxicity of pesticide residues in food;  report of a Joint Meeting of the FAO Committee on Pesticides in Agriculture and the WHO Expert Committee on Pesticide Residues.  FAO Meeting Report, No. PL:1963/13;  WHO/Food Add./23 (1964)

FAO/WHO (1965a)  —  Evaluation of the toxicity of pesticide residues in food.  Report of the Second Joint Meeting of the FAO Committee on Pesticides in Agriculture and the WHO Expert Committee on Pesticide Residues.  FAO Meeting Report, No. PL:1965/10;  WHO Food Add./26.65

FAO/WHO (1965b)  —  Evaluation of the toxicity of pesticide residues in food.  FAO Meeting Report, No. PL:1965/10/1;  WHO Food Add./27.65

FAO/WHO (1965c)  —  Evaluation of the hazards to consumers resulting from the use of fumigants in the protection of food.  FAO Meeting Report, No. PL:1965/10/2;  WHO Food Add./28.65

FAO/WHO (1967a)  —  Pesticide residues in food.  Joint report of the FAO Working Party on Pesticide Residues and the WHO Expert Committee on Pesticide Residues.  FAO Agricultural Studies, No. 73;  WHO Technical Report Series, No. 370

FAO/WHO (1967b)  —  Evaluation of some pesticide residues in food.  FAO:PL/CP/15;  WHO Food Add./67.32

FAO/WHO (1968a)  —  Pesticide residues.  Report of the 1967 Joint Meeting of the FAO Working Party and the WHO Expert Committee.  FAO Meeting Report, No. PL:1967/M/11;  WHO Technical Report Series, No. 391

FAO/WHO (1968b)  —  1967 evaluation of some pesticide residues in food.  FAO PL:1967/1967/M/11;  WHO Food Add./68.30

FAO/WHO (1969a) - Pesticide residues in food. Report of the 1968 Joint Meeting of the FAO Working Party of Experts on Pesticide Residues and the WHO Expert Committee on Pesticide Residues. FAO Agricultural Studies, No. 78; WHO Technical Report Series, No. 417

FAO/WHO (1969b) - 1968 evaluation of some pesticide residues in food. FAO PL:1968/M/9/1; WHO Food Add./69.35

FAO/WHO (1970a) - Pesticide residues in food. Report of the 1969 Joint Meeting of the FAO Working Party of Experts on Pesticide Residues and the WHO Expert Group on Pesticide Residues. FAO Agricultural Studies, No. 84; WHO Technical Report Series, No. 458

FAO/WHO (1970b) - 1969 evaluations of some pesticide residues in food. FAO PL:1969/M/17/1; WHO Food Add./70.38

FAO/WHO (1971a) - Pesticide residues in food. Report of the 1970 Joint Meeting of the FAO Working Party of Experts on Pesticide Residues and the WHO Expert Group on Pesticide Residues. FAO Agricultural Studies, No. 87; WHO Technical Report Series, No. 474

FAO/WHO (1971b) - 1970 evaluations of some pesticide residues in food. AGP:1970/M/12/1; WHO Food Add./71.42

FAO/WHO (1972) - Toxicological evaluation of certain food additives with a review of general principles and of specifications. Seventeenth report of the Joint FAO/WHO Expert Committee on Food Additives. FAO Meeting Series No. 52; WHO Technical Report Series, No. 539

FAO/WHO (1972a) - Pesticide residues in food. Report of the 1971 Joint Meeting of the FAO Working Party of Experts on Pesticide Residues and the WHO Expert Committee on Pesticide Residues. FAO Agricultural Studies, No. 88; WHO Technical Report Series, No. 502

FAO/WHO (1972b) - 1971 evaluations of some pesticide residues in food. AGP:1971/M/9/1; WHO Pesticide Residues Series, No. 1

FAO/WHO (1973a) - Pesticide residues in food. Report of the 1972 Joint Meeting of the FAO Working Party of Experts on Pesticide Residues and of the WHO Expert Committee on Pesticide Residues. FAO Agricultural Studies, No. 90; WHO Technical Report Series, No. 525

FAO/WHO (1973b) - 1972 evaluation of some pesticide residues in food. AGP:1972/M/9/1; WHO Pesticide Residues Series, No. 2

FAO/WHO (1974a) - Pesticide residues in food. Report of the 1973 Joint Meeting of the FAO Working Party of Experts on Pesticide Residues and the WHO Expert Committee on Pesticide Residues. FAO Agricultural Studies, No. 93; WHO Technical Report Series, No. 545

FAO/WHO (1974b) — 1973 evaluation of some pesticide residues in food. AGP:1973/M/9/
1; WHO Pesticide Residues Series, No. 3

FAO/WHO (1974c) — Report of the seventh session of the Codex Committee on Pesticide
Residues (document ALINORM 74/24)

FAO/WHO (1974d) — Toxicological evaluation of certain food additives with a review of
general principles and of specifications. FAO Nutrition Meeting Report Series, No.
53; WHO Technical Report Series, No. 539

FAO/WHO (1974e) — The use of mercury and alternative compounds as seed dressings.
Report of a Joint FAO/WHO Meeting. FAO Agricultural Studies, No. 95; WHO
Technical Report Series, No. 555

FAO/WHO (1975a) — Pesticide residues in food. Report of the 1974 Joint Meeting of
the FAO Working Party of Experts on Pesticide Residues and the WHO Expert Committee
on Pesticide Residues. FAO Agricultural Studies, No. 97; WHO Technical Report
Series, No. 574

FAO/WHO (1975b) — 1974 evaluations of some pesticide residues in food. AGP:1974/M/11;
WHO Pesticide Residues Series, No. 4

FAO/WHO (1975c) — Report of the eigth session of the Codex Committee on Pesticide
Residues (document ALINORM 76/24)

FAO/WHO (1976a) — Pesticide residues in food. Report of the 1975 Joint Meeting of
the FAO Working Party of Experts on Pesticide Residues and the WHO Expert Committee
on Pesticide Residues, FAO Plant Production and Protection Series, No.1; WHO
Technical Report Series, No. 592

FAO/WHO (1976b) — 1975 evaluations of some pesticide residues in food. AGP:1975/M/13;
WHO Pesticide Residues Series, No. 5

FAO/WHO (1977a) — Pesticide residues in food. Report of the 1976 Joint Meeting of the
FAO Panel of Experts on Pesticide Residues and the Environment and the WHO Expert
Group on Pesticide Residues. FAO Food and Nutrition Series, No. 9; FAO Plant
Production and Protection Series, No. 8; WHO Technical Report Series, No. 612

FAO/WHO (1977b) — 1976 evaluations of some pesticide residues in food. AGP:1977/M/14

WHO (1967) — Procedures for investigating intentional and unintentional food additives.
Report of a WHO Scientific Group. WHO Technical Report Series, No. 348

WHO (1974) — Assessment of the carcinogenicity and mutagenicity of chemicals. Report
of a WHO Scientific Group. WHO Technical Report Series, No. 546

## ANNEX 1

The following are insertions and amendments to reports of the Joint Meetings prior to the 1977 Session.

**Acephate**   Remarks: In the 1976 Report, the inclusion of <u>sheep</u> in the MRL for meat at 0.1 mg/kg should be deleted.

**Chlorpyrifos-methyl**   Remarks: This compound was incorrectly spelled "chlorpyriphos-methyl" in Annex 1 of the 1976 Report.

**Dialifos**   Remarks: In the 1976 Report, the MRL for "Fat of meat of cattle and sheep, milk (fat basis)" should be recorded as 0.2 mg/kg instead of 0.02 mg/kg.  It was correctly recorded in the Evaluations.

**Primiphos-methyl**   Add the remark "The limit for pears and plums was omitted from the Report but was correctly recorded in the Evaluations following the 1976 Meeting.

**Thiometon**   Remarks: The following MRLs were recorded incorrectly in Annex 1 to the 1976 Report:

(i)   delete "chives" in the MRLs at 0.5 mg/kg
(ii)   insert "maize (leaves, stalks, cobs)" and delete "fodder beets" in the MRLs at 0.1 mg/kg
(iii)   add new recommendation: "Fodder beets and tops, sugar beet tops 0.05 mg/kg"

ACCEPTABLE DAILY INTAKES FOR HUMANS; RESIDUE LIMITS AND GUIDELINE LEVELS
FOR COMPOUNDS CONSIDERED AT THE 1977 JOINT MEETING

The compounds listed below are those on which new data were considered at the 1977 Joint Meeting. The maximum residue limits and guideline levels include those recommended or recorded by earlier meetings and which have not been amended. New or amended ADIs and limits are indicated by (*).

Part I lists recommended maximum residue limits, temporary maximum residue limits and extraneous residue limits. MRLs and temporary MRLs apply to compounds for which an ADI for humans or temporary ADI for humans has been allocated. Where ADIs for humans or residue limits are temporary, the year in which further data are required is specified in parentheses. Extraneous residue limits are defined in the glossary in Annex 3 of the report of the 1975 meeting (FAO/WHO, 1976a). They are indicated in the Table by the symbol (E) after the recommended limit.

Part II lists guideline levels. These are derived from data relating to residues resulting from uses that are recommended or authorized by national authorities, but they are not supported by ADIs for humans (see FAO/WHO 1976a, p. 42).

Part I – ACCEPTABLE DAILY INTAKES FOR HUMANS AND MAXIMUM RESIDUE LIMITS

| Pesticide and references to previous evaluations (References are to list of FAO/WHO publications) | Recommended maximum acceptable daily intake for humans (mg/kg bw) | Commodity | Recommended maximum residue limit (mg/kg) | Remarks |
|---|---|---|---|---|
| aldrin/dieldrin 1965b, 1967b, 1968b, 1969b, 1970b, 1971b, 1975b, 1978b | 0.0001 | Potatoes | 0.2 | Limits are for aldrin or dieldrin separately or for their sum expressed as dieldrin if both are present. |
| | | Asparagus, aubergines, broccoli, Brussels sprouts, cabbage, cauliflower, cucumbers, horse radish, onions, parsnips, peppers, pimentos, radishes, radish tops | 0.1 | |
| | | Fruit | 0.05 | Extraneous residue limit for shell-free eggs is equivalent to 0.25 mg/kg in egg yolk |
| | | Rice (in husk) | 0.02 | |
| | | Carrots, lettuce, fat of meat | 0.2 (E) | |
| | | Milk and milk products (fat basis) | 0.15 (E) | |
| | | Eggs (shell-free) | 0.1 (E) | |
| | | Raw cereals (other than rice) | 0.02 (E) | |

| Pesticide and references to previous evaluations (references are to list of FAO/WHO publications) | Recommended maximum acceptable daily intake for humans (mg/kg bw) | Commodity | Recommended maximum residue limit (mg/kg) | Remarks |
|---|---|---|---|---|
| amitrole 1975b, 1978b | 0.00003 | Raw agricultural commodities of plant origin | 0.02* | Conditional ADI for humans: uses of amitrole should be restricted to those where food residues would be unlikely to occur |
| bromophos 1973b, 1976b, 1978b | 0.04 (*) | Bran (wheat) | 20 | "ADI increased and no longer temporary". |
| | | Raw grain (maize, sorghum, wheat) | 10 | |
| | | Olives, olive oil | 5 | |
| | | Apples, lamb's lettuce, leeks, radishes, white flour, whole-meal bread | 2 | |
| | | Carrots, celery, French beans, pears, plums, savoy cabbage, spinach | 1 | |
| | | (*) Blackberries, currants (red and black | 1 | |
| | | Brussels sprouts, cherries, gooseberries, lettuce, peaches, strawberries, sugarbeet (roots), white bread, fat of meat of sheep | 0.5 | |
| | | Rape seed, rape seed oil | 0.2 | |
| | | Broad beans (without pods), broccoli, cabbage (excluding savoy cabbage), cauliflower, cucumbers, kohlrabi, onions, peas | 0.1 | |
| | | Milk (whole) | 0.02* | |
| bromophos-ethyl 1973b, 1976b, 1978b | 0.003 | Fat of meat of sheep | 3 | |
| | | Apples, carrots, pears, plums, spinach, fat of meat of cattle | 2 | |
| | | Brussels sprouts, red currants | 1 | |
| | | Celeriac, cherries (sweet), gooseberries, peaches, rape-seed oil | 0.5 | |
| | | (*) Black currants, cabbage, kohlrabi, strawberries | 0.5 | |

| Pesticide and references to previous evaluations (references are to list of FAO/WHO publications) | Recommended maximum acceptable daily intake for humans (mg/kg bw) | Commodity | Recommended maximum residue limit (mg/kg) | Remarks |
|---|---|---|---|---|
| bromophos-ethyl (Cont.) | 0.003 | Lettuce, milk and milk products (fat basis) | 0.2 | |
| | | Rapeseed | 0.1 | |
| | | French beans, maize (kernels and fodder) | 0.05 | |
| | | Beans (without pod), cauliflower, milk (whole), onions, sugarbeet | 0.02* | |
| sec-butylamine 1976b, 1978b | 0.2 (1978) | Dried citrus pulp, citrus molasses | 50 | The temporary ADI for humans and maximum residue limits are for sec-butylamine, expressed as the free base |
| | | Citrus fruit | 30 | |
| | | (*) Kidney and liver of cattle, goats, pigs and sheep | 2 | |
| | | Citrus juice | 0.5 | |
| | | (*) Milk and milk products | 0.5 | |
| | | Meat of cattle, goats, pigs and sheep | 0.1 | |
| captafol 1970b, 1974b, 1975b, 1977b, 1978b | *0.1 | Apricots, peaches | 15 | Limits are for captafol only. Referred to as difolatan in FAO/WHO 1969a |
| | | Cherries (sour), pineapple (whole fruit), plums | 10 | |
| | | Cranberries, leeks | 8 | |
| | | Apples, eggplant, pears, tomatoes | 5 | |
| | | Cherries (sweet), melons (whole), peanut hulls, pumpkins | 2 | |
| | | Cucumbers (whole) | 1 | |
| | | Carrots, onions (bulb), peanuts (whole), potatoes | 0.5 | |
| | | Wheat | 0.2 | |
| | | Macadamia nuts (shelled) | 0.1 | |
| | | Peanut kernels | 0.05 | |
| captan 1965b, 1970b, 1974b 1975b, 1978b | 0.1 | Cherries | 40 | |
| | | (*) Apples, pears | 25 | |
| | | Apricots, blueberries, currants (red and black), spinach, strawberries | 20 | |
| | | Citrus fruit, endive, peaches, plums, rhubarb, tomatoes | 15 | |
| | | Cranberries, cucumbers, green beans, lettuce, peppers, raspberries | 10 | |
| | | Raisins | 5 | |

| Pesticide and references to previous evaluations (references are to list of FAO/WHO publications) | Recommended maximum acceptable daily intake for humans (mg/kg bw) | Commodity | Recommended maximum residue limit (mg/kg) | Remarks |
|---|---|---|---|---|
| carbaryl 1965b, 1967b, 1968b, 1969b, 1970b, 1971b, 1974b, 1976b, 1977b, 1978b | 0.01 | Animal feedstuffs (green), alfalfa, bean and pea vines, clover corn forage, cowpea foliage, grasses, peanut hay, sorghum forage, soybean foliage, sugar beet tops | 100 | Limits are now for carbaryl only. Those recommended in 1973 were for free and combined carbaryl, conjugated napthol, and conjugated methylol carbaryl. Limits cover residues resulting from either pre- or post-harvest uses of carbaryl |
| | | Bran | 20 | |
| | | Apricots, asparagus, blackberries, boysenberries, cherries, leafy vegetables (except brassicas), nectarines, nuts (whole), okra, olives (fresh), peaches, plums, raspberries, sorghum grain | 10 | |
| | | Blueberries, citrus fruit, cranberries, strawberries | 7 | |
| | | Apples, aubergines, bananas (pulp), barley, beans, brassicas, grapes, oats, pears, peas (incl. pod), peppers, poultry (skin), rice (in husk and hulled), rye, tomatoes, wheat | 5 | |
| | | Cucurbits (incl. melons) | 3 | |
| | | Peanuts (whole), root crops (beets, carrots, parsnips, radishes, rutabagas), wholemeal flour | 2 | |
| | | Cottonseed (whole), cowpeas, nuts (shelled), olives (processed), soybeans (dry), sweet corn (kernels) | 1 | |
| | | Eggs (shell-free), poultry (total edible portions) | 0.5 | |
| | | Meat of cattle, goats and sheep, potatoes, sugar beets, wheat flour (white) | 0.2 | |
| | | Milk, milk products | 0.1 | |

| Pesticide and references to previous evaluations (references are to list of FAO/WHO publications) | Recommended maximum acceptable daily intake for humans (mg/kg bw) | Commodity | Recommended maximum residue limit (mg/kg) | Remarks |
| --- | --- | --- | --- | --- |
| carbophenothion 1973b, 1977b, 1978b | (*) 0.0002 (1979) | Lemons | 5 | Limits are for the sum of carbophenothion, its sulphoxide and sulphone and the oxygen analogues of these compounds, expressed as carbo-phenothion |
| | | Spinach | 2 | |
| | | Fat of meat of cattle and sheep | 1 | |
| | | Apricots, grapefruit, limes, oranges, nectarines, peaches, prunes | 1 | |
| | | Apples, broccoli, Brussels sprouts, cauliflower, pears | 0.5 | |
| | | Olive oil | 0.2 | |
| | | Olives (fresh), sugar beet | 0.1 | |
| | | Milk and milk products (fat basis) | 0.1 | |
| | | Potatoes, rape seed | 0.02* | |
| | | Walnuts, pecans (shelled) | 0.02* | |
| chinomethionat 1969b, 1975b, 1978b | 0.003 (1981) | Papayas (whole fruit) | 5 | Listed in FAO/WHO 1969b as oxythioquinox, sub-sequently renamed |
| | | Citrus fruit | 0.5 | |
| | | (*) Apples | 0.2 | |
| | | Almonds (kernels), avocados, cucumbers, currants (black, red, white), gherkins, gooseberries, grapes, papaya (pulp), raw cereals | 0.1 | |
| | | Meat | 0.05* | |
| | | Macadamia nuts (kernels) | 0.02* | |
| | | Milk | 0.01* | |
| chlordane 1965b, 1968b, 1970b, 1971b, 1973b, 1975b, 1978b | 0.001 | Crude soybean and linseed oils | 0.5 | Limits are for the sum of the cis- and trans-isomers in plant products and the sum of cis- and trans-isomers and oxy-chlordane in animal products. Maximum residue limits are based on residues resulting from soil or seed treat-ments apart from that for cottonseed oil |
| | | Parsnips, potatoes, radishes, rutabagas, sugar beet, sweet potatoes, turnips | 0.3 | |
| | | Asparagus, broccoli, Brussels sprouts, cabbage, cauliflower, celery, lettuce, mustard greens, spinach, Swiss chard | 0.2 | |
| | | Almonds, bananas, cantaloups, crude cottonseed oil, cucumbers, figs, filberts, guavas, mangoes, olives, papayas, passion fruit, pecans, pineapples, pomegranates, pumpkins, squash, strawberries, walnuts, water melons | 0.1 | |
| | | Maize, oats, popcorn, rice (polished), rye, sorghum, wheat | 0.05 | |

| Pesticide and references to previous evaluations (references are to list of FAO/WHO publications) | Recommended maximum acceptable daily intake for humans (mg/kg bw) | Commodity | Recommended maximum residue limit (mg/kg) | Remarks |
|---|---|---|---|---|
| chlordane (Cont.) | 0.001 | Aubergines, beans, citrus fruit, collards (coleworts), edible cottonseed oil, edible soybean oil, peas, peppers, pimentos, pome fruit, stone fruit, tomatoes | 0.02 | |
| | | Fat of meat and poultry, milk and milk products (fat basis) | 0.05(E) | |
| | | Eggs (shell-free) | 0.02(E) | |
| chlordimeform 1972b, 1976b | 0.01 (1978) | Pears | 10 | Limits are for the sum of chlordimeform and its metabolites determined as 4-chloro-o-toluidine and expressed as chlordimeform |
| | | Peaches, prunes | 5 | |
| | | Apples, grapes, plums, straw-berries | 3 | |
| | | Brassicas, cherries, citrus fruit, cottonseed oil (crude and refined), cottonseed | 2 | |
| | | Tomatoes | 1 | |
| | | Beans | 0.5 | |
| | | Fat, meat and meat products of cattle | 0.5 | |
| | | Milk products | 0.5 | |
| | | Rice (hulled) | 0.1 | |
| | | Milk (whole) | 0.05 | |
| chlorobenzilate 1965b, 1969b, 1973b, 1978b | 0.02 | Apples | 5 | |
| | | Grapes, pears | 2 | |
| | | Cantaloups, citrus fruit, melons | 1 | |
| | | Almonds (shelled), tomatoes, walnuts (shelled) | 0.2 | |
| | | Milk (whole) | 0.05* | |
| chlorothalonil 1975b, 1978b | 0.03 (1979) | Currants (black, red, white) | 25 | |
| | | (*) Peaches | 25 | |
| | | Celery | 15 | |
| | | Blackberries, cherries, chicory sprouts, collards, endive, kale, lettuce (head), peppers, rasp-berries | 10 | |

| Pesticide and references to previous evaluations (references are to list of FAO/WHO publications) | Recommended maximum acceptable daily intake for humans (mg/kg bw) | Commodity | Recommended maximum residue limit (mg/kg) | Remarks |
|---|---|---|---|---|
| chlorothalonil (Cont.) | 0.03 (1979) | Beans (green, incl. pod), broccoli, Brussels sprouts, cabbage, cauliflower, cranberries, cucumbers, melons, onions, pumpkins, squash, tomatoes | 5 | |
| | | (*) Citrus fruit | 5 | |
| | | (*) Banana (whole) | 4 | |
| | | Carrots, sugar beet, sweet corn | 1 | |
| | | Lima beans, peanuts (whole) | 0.5 | |
| | | Peanuts (kernels), potatoes | 0.1 | |
| | | (*) Banana (pulp) | 0.1 | |
| chlorpyrifos 1973b, 1975b, 1976b, 1978b | (*) 0.001 | Fat of meat of cattle | 2 | The ADI figure has been rounded off |
| | | Apples, Chinese cabbage, grapes, kale | 1 | |
| | | Carrots, pears, tomatoes | 0.5 | |
| | | (*) Peppers | 0.5 | |
| | | Citrus fruit (whole) | 0.3 | |
| | | Aubergines, beans, raspberries, fat of meat of sheep, fat and skin of turkey | 0.2 | |
| | | Lettuce, rice (in husk), sugar beet, fat of chicken, milk and milk products (fat basis) | 0.1 | |
| | | Celery, cottonseed, cottonseed oil (crude), mushrooms, onions | 0.05 | |
| | | Cauliflower, eggs (whole), red cabbage, potatoes | 0.01* | |
| cyhexatin 1971b, 1974b, 1975b 1976b, 1978b | 0.007 (1978) | Apples, citrus fruit, pears, tea (dry, manufactured), tomatoes | 2 | Limits are for cyhexatin excluding organic degradation products and in- organic tin. Formerly listed as tricyclo- hexyltin hydroxide or as tricyclohexyl hydroxystannate |
| | | Gherkins | 1 | |
| | | Bell peppers, cucumbers, melons (glasshouse use only) | 0.5 | |
| | | Meat | 0.2(E) | |
| | | Milk and milk products (fat basis) | 0.05*(E) | |

| Pesticide and references to previous evaluations (references are to list of FAO/WHO publications) | Recommended maximum acceptable daily intake for humans (mg/kg bw) | Commodity | Recommended maximum residue limit (mg/kg) | Remarks |
|---|---|---|---|---|
| dichlofluanid 1970b, 1975b, 1978b | 0.3 (1979) | Currants (red,black,white), grapes, raspberries | 15 | Limits are for dichlofluanid only |
| | | (*)Blackberries | 15 | |
| | | Lettuce, strawberries | 10 | |
| | | (*)Gooseberries | 7 | |
| | | Apples, cucumbers, peaches, pears | 5 | |
| | | Beans (green, with pod), cherries, tomatoes | 2 | |
| | | (*)Sweet peppers | 2 | |
| | | (*)Eggplant, hops (dried) | 1 | |
| | | (*)Wheat straw | 0.5 | |
| | | (*)Onions (bulb), potatoes, raw grain (barley, oats, rye, wheat) | 0.1 | |
| dichlorvos 1965b, 1967b, 1968b, 1970b, 1971b, 1975b, 1978b | 0.004 | Cocoa beans | 5 | The limits are based on residues likely to be found at harvest or slaughter. Residues decline rapidly during storage or shipment (see FAO/WHO 1968b, 1971b). The limits for "miscellaneous food items not otherwise specified" (i.e. bread, cakes, cheese, cooked meats, etc.) is intended to cover residues resulting from usage of dichlorvos for pest control purposes in storage, in warehouses, ships, etc. |
| | | Coffee beans, lentils, peanuts, raw grain (barley, maize, oats, rice, rye, sorghum, wheat, etc.), soybeans | 2 | |
| | | Lettuce | 1 | |
| | | Milled products from raw grain, vegetables (except lettuce) | 0.5 | |
| | | Fresh fruit (apples, peaches, pears, strawberries, etc.), miscellaneous food items not otherwise specified | 0.1 | |
| | | Eggs (shell-free), meat of cattle, sheep, goats, pigs and poultry | 0.05 | |
| | | Milk (whole) | 0.02 | |
| dicloran 1975b, 1978b | (*)0.03 | Onions | 20 | The ADI is no longer temporary |
| | | Cherries, peaches | 15 | |
| | | Apricots, carrots, grapes, lettuce, nectarines, plums, raspberries, strawberries | 10 | |
| | | Blackberries, currants (black, red, white) | 5 | |
| | | Beans | 2 | |
| | | Chicory | 1 | |
| | | Gherkins, tomatoes | 0.5 | |

| Pesticide and references to previous evaluations (references are to list of FAO/WHO publications) | Recommended maximum acceptable daily intake for humans (mg/kg bw) | Commodity | Recommended maximum residue limit (mg/kg) | Remarks |
|---|---|---|---|---|
| dimethoate 1965b, 1967b, 1968b, 1971b, 1974b, 1978b | 0.02 | Blackcurrants, tree fruit (incl. citrus), vegetables (except peppers and tomatoes) | 2 | *Limits are for the sum of dimethoate and omethoate, expressed as dimethoate |
|  |  | Peppers, tomatoes | 1 |  |
|  |  | (*) Strawberries | 1 |  |
| diquat 1971b, 1973b, 1977b | (*) 0.008 | Barley, poppy seed, rice (in husk) | 5 | ADI and maximum residue limits are for diquat cation |
|  |  | Rape seed, sorghum, wheat | 2 |  |
|  |  | Cottonseed | 1 |  |
|  |  | Beans, sunflower seed | 0.5 |  |
|  |  | Potatoes, rice (polished), wheat flour | 0.2 |  |
|  |  | Maize, onions, peas, sugar beet, cottonseed oil, rapeseed oil, sesameseed oil, sunflowerseed oil | 0.1 |  |
|  |  | Other vegetable crops, meat and meat products | 0.05* |  |
|  |  | Eggs, milk (whole) | 0.01* |  |
| dithiocarbamates, dimethyl | | | | |
| ferbam, ziram thiram 1965b, 1968b, 1971b, 1975b, 1978b | (*) 0.02 0.005 (1980) | (*) Celery, currants (black, red), grapes | 5 | Limits are for residues determined and expressed as $CS_2$ |
|  |  | (*) Apples, peaches, pears, strawberries, tomatoes | 3 |  |
|  |  | (*) Bananas (whole), cherries, endive, lettuce, melons, plums | 1 |  |
|  |  | (*) Beans (in pod), carrots, cucumbers | 0.5 |  |
|  |  | (*) Wheat | 0.2 |  |
|  |  | (*) Bananas (pulp), potatoes | 0.1 |  |

| Pesticide and references to previous evaluations (references are to list of FAO/WHO publications) | Recommended maximum acceptable daily intake for humans (mg/kg bw) | Commodity | Recommended maximum residue limit (mg/kg) | Remarks |
|---|---|---|---|---|
| dithiocarbamates ethylenebis<br><br>(mancozeb, maneb and zineb, including zineb derived from nabam plus zinc sulfate)<br>1965b, 1968b, 1971b, 1975b, 1978b | 0.005 (1980) | (See Remarks) | | Recommended maximum residue limits are the same as for dithiocarbamates dimethyl. They are for residues determined and expressed as $CS_2$.<br><br>Guideline levels for ethylenethiourea (ETU) are recorded in Part II. |
| dithiocarbamates propylenebis (propineb) 1978b | *0.005 (1980) | (See Remarks) | | Recommended maximum residue limits are the same as for dithiocarbamates, dimethyl. They are for residues determined and expressed as $CS_2$ |
| dodine 1975b, 1977b, 1978b | 0.01 | Grapes, peaches, strawberries<br>(*) Apples, pears<br>Cherries | 5<br>5<br>2 | |
| ethiofencarb 1978b | 0.1 (*) (1978) | (*) Cherries, lettuce<br>(*) Apples, apricots, artichokes, beans with pod, beet tops, brassicas, peaches, pears, plums, sugar beet tops (leaves)<br>(*) Currants (black, red), egg-plants, wheat straw<br>(*) Cucumbers<br>(*) Potatoes, radishes<br>(*) Beans without pod, soybeans without pod<br>(*) Fodder beet and sugar beet (roots)<br>(*) Raw grain (barley, oats, rye, wheat)<br>(*) Eggs, milk, meat of cattle, pigs and poultry | 10<br><br><br>5<br><br>2<br>1<br>0.5<br><br>0.2<br><br>0.1<br><br>0.05<br><br>0.02* | Limits are for the sum of ethiofencarb, its sulphoxide and its sulphone, expressed as ethiofencarb |

| Pesticide and references to previous evaluations (references are to list of FAO/WHO publications) | Recommended maximum acceptable daily intake for humans (mg/kg bw) | Commodity | Recommended maximum residue limit (mg/kg) | Remarks |
|---|---|---|---|---|
| fenamiphos 1975b, 1978b | 0.0006 | (*) Oranges (whole fruit)<br>Tomatoes<br>(*) Potatoes<br>Bananas, coffee beans (green and roasted), grapes, sweet potatoes<br>(*) Oranges (flesh of fruit)<br>Broccoli, Brussels sprouts, cabbage, carrots, cauliflower, cottonseed, melons, peanuts (kernels), pineapple, soybeans (dried), sugar beets<br>(*) Citrus fruit (other than oranges) (whole fruit) | 0.5<br>0.2 (1980)<br>0.2<br>0.1<br>0.1<br><br>0.05*<br>0.05* (1980) | Limits are for the sum of fenamiphos, its sulphoxide and its sulphone, expressed as fenamiphos. The limits for citrus fruit and tomatoes are temporary |
| fenbutatin oxide 1978b | (*) 0.03 | (*) Dried apple pomace<br>(*) Peaches, dried citrus pulp<br>(*) Apples, cherries (sour and sweet), citrus fruit, pears<br>(*) Plums, strawberries<br><br>(*)Cucumbers, eggplants, gherkins, melons, sweet peppers, tomatoes<br>(*) Liver and kidney of cattle, goats, horses, pigs and sheep<br>(*) Meat of cattle, goats, horses, pigs and sheep<br>(*) Milk | 20<br>7<br>5<br>3<br><br>1<br><br>0.2<br><br>0.02*<br>0.02* | Limits are for fenbutatin oxide only |
| fenitrothion 1970b, 1975b, 1977b 1978b | 0.005 | Bran of rice and wheat<br>(*) Raw cereals<br>Wheat flour (wholemeal)<br>Peaches<br>Wheat flour (white), rice (milled)<br>Apples, cabbage, red cabbage, cherries, grapes, lettuce, peas, strawberries, tea (green, dry), tomatoes<br>Bread (white), leeks, oranges, radishes<br>Cauliflower, cocoa beans, eggplant, pears, peppers, soybeans (dry)<br>Cucumbers, meat, fat of meat, milk, milk products, onions, potatoes | 20<br>10<br>5<br>2<br>1<br><br>0.5<br><br>0.2<br><br>0.1<br>0.05* | Limits are for fenitrothion and its oxygen analogue, expressed as fenitrothion. "Raw cereals" replaces "Rice (in husk and hulled), wheat". |

| Pesticide and references to previous evaluations (references are to list of FAO/WHO publications) | Recommended maximum acceptable daily intake for humans (mg/kg bw) | Commodity | Recommended maximum residue limit (mg/kg) | Remarks |
|---|---|---|---|---|
| fenthion 1972b, 1976b, 1978b | 0.0005 (1978) | Apples, cherries, lettuce, peaches, fat of meat | 2 | Limits are for the sum of fenthion, its sulphoxide, its sulphone and the oxygen analogues of these compounds determined as the sulphone of the oxygen analogue and expressed as fenthion. (The temporary nature of ADI erroneously omitted from Annex I of the Report on 1975 Meeting, p.27) |
| | | (*) Citrus, pears, strawberries | 2 | |
| | | Cabbage, cauliflower, olives, olive oil | 1 | |
| | | (*) Bananas, plums | 1 | |
| | | Grapes, peas, meat | 0.5 | |
| | | (*) Tomatoes | 0.5 | |
| | | Squash | 0.2 | |
| | | (*) Citrus juice | 0.2 | |
| | | Rice, wheat, milk products (fat basis) | 0.1 | |
| | | (*) Beans, onions, sweet potatoes | 0.1 | |
| | | Milk (whole) | 0.05 | |
| | | (*) Potatoes | 0.05* | |
| ferbam 1965b, 1968b, 1971b, 1975b, 1978b | 0.02 | | | See dithiocarbamates, dimethyl |
| imazalil 1978b | (*) 0.01 (1981) | (*) Citrus fruit (whole) | 5 | Limits are for imazalil only |
| | | (*) Bananas (whole) | 2 | |
| | | (*) Cucumbers, gherkins | 0.5 | |
| | | (*) Bananas (without peel) | 0.2 | |
| | | (*) Citrus fruit (without peel), wheat straw | 0.1 | |
| | | (*) Wheat grain | 0.01* | |
| iprodione 1978b | (*) 0.3 | (*) Apples, grapes, lettuce, peaches, pears | 10 | Limits are for iprodione only |
| | | (*) Plums, strawberries | 7 | |
| | | (*) Blackcurrants, cucumbers, raspberries, sweet peppers, tomatoes | 5 | |
| | | (*) Rice (husked, unpolished) | 3 | |
| | | (*) Chicory (witloof), sprouts | 1 | |
| | | (*) Beans (dry) | 0.2 | |
| | | (*) Garlic, onions | 0.1 | |

| Pesticide and references to previous evaluations (references are to list of FAO/WHO publications) | Recommended maximum acceptable daily intake for humans (mg/kg bw) | Commodity | Recommended maximum residue limit (mg/kg) | Remarks |
|---|---|---|---|---|
| lindane 1965b, 1967b, 1968b, 1969b, 1970b, 1972b, 1974b, 1975b, 1976b, 1978b | (*) 0.01 | Cranberries, strawberries | 3 | Referred to as gamma-BHC prior to 1964. ADI no longer temporary |
| | | Endive, lettuce, fat of meat of cattle, pigs and sheep | 2 | |
| | | (*) Spinach | 2 | |
| | | Beans (dried), cocoa beans, kohlrabi, radish | 1 | |
| | | Apples, Brussels sprouts, cabbage, cauliflower, cherries, grapes, pears, plums, raw cereals, red currants, rice (in husk), savoy cabbage | 0.5 | |
| | | (*) Tomatoes | 0.5 | |
| | | Peas, sugar beet (roots), sugar beet (leaves) | 0.1 | |
| | | Potatoes, rape seed | 0.05* | |
| | | Poultry (fat basis) | 0.7(E) | |
| | | Eggs (shell-free), milk and milk products (fat basis) | 0.1(E) | |
| | | (*) Carrots | 0.2(E) | |
| malathion 1965b, 1967b, 1968b, 1969b, 1970b, 1971b, 1974b, 1976b, 1978b | 0.02 | (*) Bran of rye and wheat | 20 | Limits are for the sum of malathion and its oxygen analogue expressed as malathion |
| | | Blackberries, cabbage, dried fruit, endive, grapes, lettuce, pulses, (dried beans, lentils), nuts, raspberries, raw cereals, spinach | 8 | |
| | | Cherries, peaches, plums | 6 | |
| | | Broccoli | 5 | |
| | | Citrus fruit | 4 | |
| | | Kale, tomatoes, turnips | 3 | |
| | | Apples, beans (green), wholemeal and flour from rye and wheat | 2 | |
| | | Celery, strawberries | 1 | |
| | | Aubergines, blueberries, cauliflower, collards, kohlrabi, pears, peas (in pod), peppers, root vegetables (except turnips), Swiss chard | 0.5 | |
| mancozeb 1968b, 1971b, 1975a, 1978b | 0.005 (1980) | | | See dithiocarbamates, ethylenebis |

| Pesticide and references to previous evaluations (references are to list of FAO/WHO publications) | Recommended maximum acceptable daily intake for humans (mg/kg bw) | Commodity | Recommended maximum residue limit (mg/kg) | Remarks |
|---|---|---|---|---|
| maneb<br>1965b, 1968b, 1971b, 1975a, 1978b | 0.005<br>(1980) | | | See dithiocarbamates, ethylenebis |
| methoxychlor<br>1965b, 1978b | 0.1 | | | Previously established ADI unchanged |
| pirimiphos-methyl<br>1975b, 1977b, 1978b | 0.01 | Peanuts (whole), peanut hulls | 50 | Previous recommendations for individual cereals replaced by single limit for "raw cereals" |
| | | Bran (wheat, rice) | 20 | |
| | | Peanut oil, raw cereals (except rice, hulled or polished) | 10 | |
| | | Lettuce, mushrooms, olives, peanuts (kernels), spinach, wholemeal flour (wheat, rye) | 5 | |
| | | Apples, Brussels sprouts, cabbage, cauliflower, cherries, pears, plums | 2 | |
| | | Blackcurrants, carrots, cucumbers, gooseberries, peppers, raspberries, spring onions, strawberries, tomatoes | 1 | |
| | | Beans (with pod), cheese, citrus, dates | 0.5 | |
| | | Peas, potatoes | 0.05* | |
| propargite<br>1978b | (*) 0.08<br>(1981) | (*) Apple pomace | 80 | |
| | | (*) Alfalfa hay | 75 | |
| | | (*) Almond hulls | 55 | |
| | | (*) Alfalfa (fresh), mint hay | 50 | |
| | | (*) Dried citrus pulp, grape pomace | 40 | |
| | | (*) Dried hops | 30 | |
| | | (*) Raisins | 25 | |
| | | (*) Beans (in pod) | 20 | |
| | | (*) Corn fodder and forage, cranberries, grapes, peanut hay and forage, sorghum fodder & forage | 10 | |
| | | (*) Apricots, nectarines, peaches, plums, strawberries | 7 | |
| | | (*) Citrus, sorghum grain | 5 | |
| | | (*) Apples, figs, pears | 3 | |
| | | (*) Milk (fat) | 2 | |
| | | (*) Beans (dry) | 0.2 | |

| Pesticide and references to previous evaluations (references are to list of FAO/WHO publications) | Recommended maximum acceptable daily intake for humans (mg/kg bw) | Commodity | Recommended maximum residue limit (mg/kg) | Remarks |
|---|---|---|---|---|
| propargite (Cont.) | (*) 0.08 (1981) | (*) Almonds, cottonseed, maize (kernels), peanuts (kernels), potatoes, walnuts, eggs, fat of meat and poultry, milk (whole) | 0.1 | |
| propineb 1978b | (*) 0.005 (1980) | | | See dithiocarbamates, propylenebis |
| propoxur 1974b, 1978b | 0.02 | Animal feedstuffs (green) | 5 | (*) Limits are for the sum of propoxur and its main metabolites, i.e. 2-hydroxyphenyl methylcarbamate and 2-isopropoxyphenyl hydroxymethylcarbamate, expressed as propoxur |
| | | Apples, cherries, peaches, pears, plums | 3 | |
| | | Blackberries, gooseberries, red currants, strawberries | 3 | |
| | | Vegetables (except potatoes and root vegetables) | 3 | |
| | | Potatoes, root vegetables | 0.5 | |
| | | Rice grain (rough) | 0.5 | |
| | | Rice (hulled) | 0.1 | |
| | | Cocoa beans, meat, milk | 0.05* | |
| quintozene 1970b, 1974b, 1975b, 1976b, 1978b | 0.007 | Peanuts (whole) | 5 | (*) Limits are for the sum of quintozene, pentachloroaniline and methyl pentachlorophenyl sulphide |
| | | Lettuce | 3 | |
| | | Peanuts (kernels) | 2 | |
| | | Bananas (whole) | 1 | |
| | | Beans (navy), potatoes | 0.2 | |
| | | Tomatoes | 0.1 | |
| | | Cottonseed | 0.03 | |
| | | Broccoli, cabbage | 0.02 | |
| | | Bananas (pulp), beans (other than navy), peppers (bell) | 0.01 | |
| thiabendazole 1971b, 1972b, 1973b, 1976b, 1978b | (*) 0.3 | Apples, citrus fruit, pears | 10 | Limits are for thiabendazole in food of plant origin and for the sum of thiadendazole and 5-hydroxythiabendazole in food of animal origin. ADI has been increaed. |
| | | (*) Sugar beet tops | 10 | |
| | | (*) Sugar beet, sugar beet pulp, potatoes (washed before analysis) | 5 | |
| | | Bananas | 3 | |
| | | (*) Sugar beet molasses | 1 | |
| | | Bananas (pulp) | 0.4 | |
| | | (*) Raw grain | 0.2 | |
| | | Meat & meat products of cattle, goats, horses, pigs and sheep, milk | 0.1* | |
| | | (*) Onions, strawberries, tomatoes | 0.1 | |

| Pesticide and references to previous evaluations (references are to list of FAO/WHO publications) | Recommended maximum acceptable daily intake for humans (mg/kg bw) | Commodity | Recommended maximum residue limit (mg/kg) | Remarks |
|---|---|---|---|---|
| thiophanate-methyl 1974b, 1976b, 1978b | 0.08 | Celery | 20 | Limits are for the sum of thiophanate-methyl and carbendazim, expressed as carbendazim |
| | | Cherries, citrus fruit, grapes, peaches, raspberries | 10 | |
| | | Apples, pears, blackcurrants, gooseberries, strawberries, carrots, lettuce, tomatoes, sugar beet (tops) | 5 | |
| | | Plums, beans (broad, dwarf, French, runner, kidney), gherkins | 2 | |
| | | Bananas (whole), mushrooms | 1 | |
| | | Cucumber | 0.5 | |
| | | Raw cereals, onions, sugar beets | 0.1* | |
| | | Meat and fat of chickens | 0.02* | |
| thiram 1965b, 1968b, 1975a, 1978b | 0.005 | | | See dithiocarbamates, dimethyl |
| zineb 1965b, 1968b, 1971b, 1975a, 1978b | 0.005 (1980) | | | See dithiocarbamates, ethylenebis |
| ziram 1965b, 1968b, 1971b, 1975a, 1978b | 0.02 | | | See dithiocarbamates, dimethyl |

Part II — GUIDELINE LEVELS (GL) (NO ACCEPTABLE DAILY INTAKES)

| Pesticide | | Commodity | Level | |
|---|---|---|---|---|
| carbendazim 1974b, 1977b, 1978b | | Cherries, citrus fruit, grapes peaches | 10 (GL) | |
| | | Apples, gooseberries, lettuce, pears, strawberries, sugar beet tops, tomatoes | 5 (GL) | |
| | | Beans (dwarf), celery, gherkins, peanut hay, plums, wheat straw | 2 (GL) | |
| | | Bananas (whole), mushrooms | 1 (GL) | |
| | | Bananas (pulp), cucumbers, melons | 0.5 (GL) | |
| | | Peanut hulls | 0.2 (GL) | |
| | | Coffee beans (raw), peanuts, raw cereals, sugar beet | 0.1* (GL) | |

| Pesticide and references to previous evaluations (references are to list of FAO/WHO publications) | Commodity | Guideline level (mg/kg) | Remarks |
|---|---|---|---|
| daminozide 1978b | (*) Sour cherries | 60 | |
| | (*) Plums | 50 | |
| | (*) Tomatoes | 40 | |
| | (*) Apples, nectarines, peaches, peanut (kernels), pears, sweet cherries | 30 | |
| | (*) Brussels sprouts | 20 | |
| | (*) Grapes, peanut hay | 10 | |
| | (*) Melons | 3 | |
| | (*) Peppers | 1 | |
| | (*) Eggs, meat | 0.2* | |
| | (*) Milk | 0.05* | |
| ethephon 1978b | (*) Blackberries, peppers | 30 | |
| | (*) Blueberries | 20 | |
| | (*) Cherries | 10 | |
| | (*) Apples, blackcurrants, cranberries, figs | 5 | |
| | (*) Tomatoes | 3 | |
| | (*) Cantaloups, lemons, pineapples | 2 | |
| | (*) Filberts, tangerines, walnuts | 0.5 | |
| | (*) Coffee (beans) | 0.1 | |
| ethylene thiourea (ETU) 1975a | Beans (in pod) | 0.1 | See also dithiocarbamates, ethylenebis |
| | Tomatoes | 0.05 | |
| | Apples, pears | 0.02 | |
| | Banana (pulp), carrots, celery, lettuce, maize, sweet corn (cob and kernels, husks and silks removed), potatoes | 0.01* | |
| maleic hydrazide 1977b, 1978b | (*) Tobacco leaf | 100 | Guideline levels are for the sum of free and bound unchanged maleic hydrazide and its $\beta$-D-glucoside |
| | Potatoes | 50 | |
| | (*) Tobacco products (cigarettes, cigars, smoking tobacco, chewing tobacco, snuff) | 50 | |
| | Onions | 15 | |

| Pesticide and references to previous evaluations (references are to list of FAO/WHO publications) | Commodity | Guideline level (mg/kg) | Remarks |
|---|---|---|---|
| methomyl 1976b, 1977b, 1978b | Alfalfa, barley straw, oat straw, pea forage, sorghum forage, soybean forage, wheat straw | 10 | |
| | Cabbage, collards, grapes, lettuce, nectarines, peas (incl. pods), peaches, peanut forage, spinach | 5 | |
| | Celery | 3 | |
| | Apples, asparagus, citrus (oranges, lemons, grapefruit, tangelos), mint hay, snap beans | 2 | |
| | Cauliflower, grapes, hops (dried), peppers, tobacco (flue cured), tomatoes | 1 | |
| | Cucumbers, eggplant, onions (green), peas (shelled, excluding pods) | 0.5 | |
| | Barley, cantaloups, melons, onions (dry), pineapples, sorghum, summer squash, water melons | 0.2 | |
| | Beans (dry), cottonseed, oats, peanuts, peanut hulls, potatoes, soybeans, sugar beets, sweet corn, wheat | 0.1 | |
| | Meat, milk | 0.02* | |
| nabam 1965b, 1966b, 1978b | | | ADI for humans was withdrawn |
| phorate 1978b | (*) Alfalfa (dry), sugar beet tops | 1 | Limits are for the sum of phorate, its sulphoxide, its sulphone and the oxygen analogues of these compounds, determined as the sulphone of the oxygen analogue and expressed as phorate |
| | (*) Carrots | 0.5 | |
| | (*) Lettuce | 0.2 | |
| | (*) Beans, celery, cowpea, eggplant, hops (dried), rape seed, tomatoes | 0.1 | |
| | (*) Barley, cottonseed, grapes, maize (green), peanuts (without shell), potatoes, sorghum, soybeans, sugar beet, fodder beet, wheat | 0.05 | |
| | (*) Eggs, meat, milk | 0.05* | |

* Level at or about the limit of determination.

ANNEX 2

FURTHER WORK OR INFORMATION REQUIRED

(OR DESIRABLE)

If a compound has been considered at earlier meetings, the requirements listed below replace those stated in earlier reports, unless otherwise indicated.

ALDRIN/DIELDRIN

Desirable

1.   Further studies on the possible mechanism of the tumorigenic action on mouse liver.

AMITROLE

Desirable

See FAO/WHO, 1975a, p.31

BROMOPHOS

Desirable

1.   Further information on residues in rice following storage and processing under full-scale commercial conditions.

2.   Further information on the levels and fate of bromophos residues in products of animal origin

BROMOPHOS-ETHYL

Desirable

See Report of 1975 Meeting (FAO/WHO 1976a, pp. 31-32)

sec-BUTYLAMINE

Required (before 30 June, 1978)

1.   Quantitative metabolic studies in animals

Desirable

1.   Mutagenicity studies with techniques currently available.

2.   Clinical observations in man.

3.   Information on the level and fate of sec-butylamine residues in potatoes.

CAPTAFOL

Desirable

1.   Further studies to investigate the metabolism of the tetrachloroethylthio moiety of captafol.

2. Additional information on national use patterns and data from supervised residue trials, particularly on apples and pears, from countries where the product is used.

3. Further data on the effects of washing, peeling and blanching on residue levels in various crops.

4. Information, some of it from studies now in progress, on new methods for the analysis of the parent compound together with the main metabolites in products of animal origin.

5. Results of studies now in progress on the level and nature of captafol residues in meat, milk, poultry and eggs.

6. Results of studies now in progress on the fate of captafol residues in citrus fruits and citrus pulp.

CAPTAN

## Desirable

1. Investigation of the significance of haematoma formation in the foetus in relation to foetal death and malformation.

2. Details of recent studies on mutagenicity and carcinogenicity mentioned in the Report of the 1977 Meeting (FAO/WHO, 1978a), Section 4.7, "Toxicology".

3. Information on current national use patterns and corresponding supervised residue trials.

CARBARYL

See Report of 1976 Meeting (FAO/WHO 1977a, p. 29)

CARBENDAZIM

See Report of 1976 Meeting (FAO/WHO 1977a, p. 27)

CARBOPHENOTHION

## Required (before July 1979)

1. Further studies to substantiate the marked species difference in sensitivity to plasma cholinesterase depression.

## Desirable

1. Studies on the identity and relative toxicity of metabolites.

2. An additional carcinogenicity study in another species, in view of the hepatic toxicity observed in rodents.

CHLORDANE

## Desirable

1. Further studies to elucidate the long-term effects.

2. Further information relating to the occurrence of residues in various commodities as listed by the Codex Committee on Pesticide Residues at its Ninth Session (1977, ALINORM 78/24, p. 10)

## CHLORDIMEFORM

<u>Required</u> (before July 1978) (in addition to further work listed in FAO/WHO 1976a, pp.32-33)

1.    Results on long-term studies with 4-chloro-o-toluidine.

## CHLOROBENZILATE

<u>Desirable</u>

1.    Information on the level and fate of chlorobenzilate residues in tea.

## CHLOROTHALONIL

<u>Required</u> (before July 1979)

1.    Additional studies to resolve the lower dose limits for kidney effects in rats.

2.    Studies to define the growth reduction after administration of chlorothalonil or its metabolite 4-hydroxy-2,5,6-trichloroisothalonile in pups relative to ingestion or secretion into milk.

<u>Desirable</u>

1.    Observation in humans.

2.    Information on the extent of metabolism to 4-hydroxy-2,5,6-trichloroisothalonile in mammals.

3.    Information on the effects of cooking on residues.

## CHLORPYRIFOS

<u>Desirable</u>

1.    Elucidation of possible increased sensitivity to plasma cholinesterase depression after withdrawal from an initial dose regime.

## CYHEXATIN

<u>Required</u> (before July 1978)

See report on 1973 Meeting (FAO/WHO 1874a, pp. 41-42).  Then referred to as tricyclohexyltin hydroxide.

## DAMINOZIDE

<u>Required</u> (before an acceptable daily intake for humans (ADI) and maximum limits (MRL) can be established)

1.    An adequate long-term rat study.

2.    Information on biotransformation in animals.

<u>Desirable</u>

1.    An analytical method more suitable for regulatory purposes than the present colorimetric method.

2.  Information on the occurrence of residues in foods in commerce.

3.  Further identifiction of the terminal residues in crops and in foods of animal origin.

DICHLOFLUANID

Required (by July 1979)

1.  Studies to elucidate the accumulation seen in the thyroid.

2.  Results from current studies on the pathways of degradation, especially the fate of the fluorine-containing moiety of the molecule in or on plants.

Desirable

1.  Further residue data on dichlofuanid residues (parent compound) resulting from supervised trials especially on melons and leaf brassicas.

DICLORAN

Desirable

1.  Further observation in humans.

2.  Results of further studies of the level and fate of dicloran residues on apricots and nectarines, particularly following post-harvest treatment.

DITHIOCARBAMATES

(a)  Dithiocarbamates (in general)

Required (by July 1980)

1.  A comprehensive survey of the use patterns of thiram, ferbam and ziram.

2.  Data on residues of thiram, ferbam and ziram from supervised trials.

3.  Data on residues of dithiocarbamates from crops grown under glass.

(b)  Dimethyl dithiocarbamates

THIRAM

Required (by July 1980)

Further studies to investigate the haematological effects.

(c)  Ethylene bis dithiocarbamate

MANEB, MACOZEB, ZINEB

Required (by July 1980) (in addition to further work listed in FAO/WHO 1975a, pp.34-35)

1.  Further data on the conversion of EBDC to ETU in various food processing procedures.

2.  Studies of procedures to minimize the formation of ETU during food processing.

(d)  Propylene bis dithiocarbamates.

PROPINEB

<u>Required</u> (by July 1980)

1. A study on the mechanism of action of propineb on the thyroid.

2. Further work to elucidate the mode of action of PTU and its long-term toxicity.

3. Information regarding the fate of residues during food processing, including cooking.

DODINE

<u>Desirable</u>

1. A feeding study with large animals to determine whether feeding apple pomace and grapes contribute residues to meat and milk.

ETHEPHON

<u>Required</u> (before an acceptable daily intkae for humans (ADI) can be established and maximum residue limits (MRL) can be recommended)

1. Submission of full toxicological data.

<u>Desirable</u>

1. Country statements on national use patterns, supervised residue trials and information on occurrence of residues on commodities in commerce.

ETHIOFENCARB

<u>Required</u> (by July 1978)

1. Submission of the results of the reproduction study now in progress.

2. Further residue data from supervised trials on apricots, artichokes, various brassica, cherries, cucumbers (including those grown under glass), eggplants, plums, radishes, soybeans.

<u>Desirable</u>

1. Further residue data from supervised trials on other crops on which ethiofencarb is known to be used.

FENAMIPHOS

<u>Required</u> (by 1980)

1. Further data on which to assess the residues in or on citrus fruits other than oranges and on tomatoes.

<u>Desirable</u>

1. Information on brain cholinesterase and behavioural studies in animals exposed to low levels for extended periods.

2. Observations in man.

3.   Additional studies on the potentiation effects with other organophosphorus pesticides.

4.   Residue data for raw agricultural products moving in commerce.

5.   Further residue data for different crops from supervised trials, especially for beans, cucumbers, lettuce, peppers and strawberries.

## FENBUTATIN OXIDE

### Desirable

1.   Further studies to elucidate the EEG findings in the dog.

2.   Elucidation of species differences in toxic effects on the rabbit testes.

## FENITROTHION

### Desirable

1.   Further observations in humans.

2.   Information on the level and fate of fenitrothion residues on citrus varieties other than oranges.

## FENTHION

### Required (by July 1978)

1.   Adequate two-year feeding studies in the dog and in one rodent species.

2.   Establishement of the sequence of metabolic changes in man and laboratory animals in order to elucidate the mechanism of long-lasting cholinesterase inhibition.

### Desirable

1.   Information from additional countries on crops already considered and on additional crops in order to confirm the level of residues following approved uses of fenthion.

2.   Information on the effect of processing and cooking on fenthion residues in fruits and vegetables.

## IMAZALIL

### Required (by July 1981)

1.   Comparative acute toxicity studies of the different salts.

2.   Further long-term studies.

### Desirable

1.   Pharmacokinetic studies of the different salts.

2.   Metabolic studies with imazalil labelled with $^{14}C$.

3.   Further information on the nature and level of metabolites and degradation products in plants.

## IPRODIONE

<u>Desirable</u>

1.  Information on the fate of residues in milk, meat and eggs when food wastes containing iprodione residues are used as components of animal feeds.

2.  Residue data on grain and straw from supervised trials on cereal crops treated according to good agricultural practice.

3.  Further information about the effects or processing and cooking on iprodione residues in a range of commodities.

## LINDANE

<u>Desirable</u>

1.  Additional residue data, especially from countries outside Europe in which lindane is used.

## MALATHION

<u>Required</u>

1.  Further residue data from supervised trials of malathion on stored grains including data on the fate of malathion during milling, baking and cooking.

## MALEIC HYDRAZIDE

<u>Required</u> (before an acceptable daily intake for humans (ADI) can be established and maximum residue limits (MRL) can be recommended).

"See Report of 1976 Meeting (FAO/WHO, 1977a, Annex 2).  Data on the effects of cooking on residues in potatoes and on the carry-over of maleic hydrazide from raw into cured tobacco and into cigarette smoke are no longer required.  The other requirements remain."

## METHOMYL

<u>Required</u> (before an acceptable daily intake for humans (ADI) can be established and maximum residue limits (MRL) can be recommended).

1.  Submission of full toxicological data.

<u>Desirable</u> (As in Report of 1976 Meeting (FAO/WHO 1977a, p. 33)

## PHORATE

<u>Required</u> (before an acceptable daily intake for humans (ADI) can be established and maximum residue limits (MRL) can be recommended).

1.  Long-term carcinogenicity studies

2.  Teratogenicity studies

3.  Studies on potential neurotoxicity

4.  Studies on the toxicity of metabolites

<u>Desirable</u>

1.   Observation in humans.

PHOSMET

<u>Required</u>

See Report on 1976 Meeting (FAO/WHO, 1977a, p. 34)

PIRIMIPHOS-METHYL

<u>Desirable</u> (in addition to further work listed in FAO/WHO 1977a, p. 35)

1.   Further information on the level and fate of residues in food at the point of consumption following the use of primiphos-methyl for the control of various stored pproduct pests.

PROPARGITE

<u>Required</u> (before July 1981)

1.   A carcinogenicity study

<u>Desirable</u>

1.   Mutagenicity studies

2.   Observations in humans

3.   Data from supervised trials in countries other than the U.S.A.

4.   Information on the occurrence of residues on commodities in commerce.

5.   Confirmation of the postulated degradation pathway in plants using radio-labelled propargite.

THIABENDAZOLE

<u>Desirable</u>

1.   Information on the use pattern on cereals and an indication as to whether the residue pattern is similar on rice, barley and oats to that on wheat.

TRIFORINE

<u>Required</u> (by 30 June 1978 and before an acceptable daily intake for humans (ADI) can be established and maximum residue limits (MRL) can be recommended).

1.   Submission of full data relating to toxicological and residue studies.